信息技术及应用英语教程

English for Information Technology and Application

主　编　李玉华　魏　薇

副主编　周香花　李　力　谭　添　邱心莹

编　者　邹丽琴
　　　　支菊芳　贾淑华
　　　　戴晓玲　熊艳茹

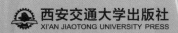

西安交通大学出版社
XI'AN JIAOTONG UNIVERSITY PRESS

图书在版编目（CIP）数据

信息技术及应用英语教程：英文/李玉华，魏薇主编．—西安：
西安交通大学出版社，2023.8
　ISBN 978 - 7 - 5693 - 3329 - 9

　Ⅰ．①信…　Ⅱ．①李…　②魏…　Ⅲ．①电子计算机-英语-
教材　Ⅳ．①TP3

中国国家版本馆 CIP 数据核字（2023）第 122681 号

信息技术及应用英语教程

XINXI JISHU JI YINGYONG YINGYU JIAOCHENG

主　　编	李玉华　魏　薇
责任编辑	李　蕊
责任校对	庞钧颖
封面设计	任加盟

出版发行	西安交通大学出版社
	（西安市兴庆南路 1 号　邮政编码 710048）
网　　址	http://www.xjtupress.com
电　　话	（029）82668357　82667874（市场营销中心）
	（029）82668315（总编办）
传　　真	（029）82668280
印　　刷	陕西奇彩印务有限责任公司

开　　本	787mm×1092mm　1/16　　**印张**　9.75　　**字数**　291 千字
版次印次	2023 年 8 月第 1 版　　2023 年 8 月第 1 次印刷
书　　号	ISBN 978 - 7 - 5693 - 3329 - 9
定　　价	39.80 元

前　言

《大学英语教学指南（2020 版）》指出，大学英语的教学目标及培养方向应在注重发展学生通用语言能力的同时，进一步增强其学术英语或职业英语交流能力和跨文化交际能力，以便学生在日常生活、专业学习和就业从业等不同领域或语境中能够用英语进行有效交流。为了使学生适应大学英语改革后的学习要求，不断提升专门用途英语能力，我们特编写了针对计算机、软件工程、电子信息工程、物联网、通信工程、电气工程等相关专业的专门用途英语教材——《信息技术及应用英语教程》（以下简称《教程》）。

《教程》充分体现了专门用途英语的特色，所选主题反映了行业发展的主线脉络，所选文章题材新颖，从概念介绍到实际应用都紧跟时代发展，既有知识性，又有趣味性，内容完善，符合学生的学习需求。《教程》的各单元结构设计合理，从单元主题、词汇到拓展技能都做到了新颖编排与实用性的结合，课后练习也丰富多样。

《教程》共十个单元，各单元主题分别为：计算机的历史与发展、网络与应用、软件开发、计算机安全、电子商务、计算机辅助学习、社交媒体、云计算、人工智能、虚拟现实。每个单元分为四个部分。第一部分通过术语匹配、图片识别及选词填空练习来介绍和学习相关专业术语及重点词汇，扫除专业词汇和重点词汇障碍。第二部分包括两篇阅读文章：A 篇侧重于概念的介绍，加深学生对技术本身的了解；B 篇聚焦技术在各领域的应用，贴近学生生活，便于理解和应用。第二部分的两篇课文均附有适量阅读理解练习，便于学生集中训练阅读技能，提升阅读和理解能力。第三部分的重点在于拓展技能，通过各种不同形式的训练提升学生的各项基本技能，为日后求职奠定基础。第四部分为扩展练习，涵盖了与该单元主题相关的英译中、中译英和写作练习，强化学生在专业领域综合运用英语的能力。

《教程》是南昌大学科学技术学院 ESP 课程"信工英语"的指定教材，该书的出版得到了南昌大学科学技术学院和西安交通大学出版社的大力支持，在此一并致谢。限于编者的水平，书中错误及不足之处在所难免，望广大读者批评指正。

编者

2023 年 4 月

Contents

Unit 1

The History of Computers and Internet

Part 1   Vocabulary

A. Choose the full names in Column B that best match the terms in Column A.

Column A	Column B
1. ARPANET	a. Universal Automatic Computer
2. TCP	b. World Wide Web
3. IP	c. Advanced Research Project Agency Network
4. LAN	d. Peer to Peer
5. HTML	e. Transmission Control Protocol
6. URL	f. Wireless Fidelity
7. UNIVAC	g. Hypertext Markup Language
8. Wi-Fi	h. Local Area Network
9. WWW	i. Uniform Resource Locator
10. P2P	j. Internet Protocol

B. Use the words from Exercise A to label the following pictures.

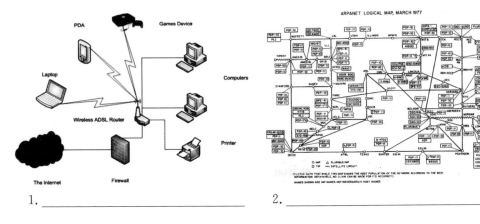

1. _____   2. _____

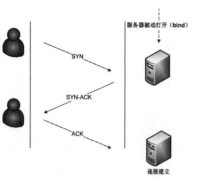

3. _____   4. _____

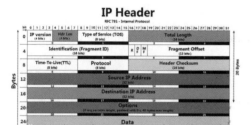

5. _____   6. _____

C. Complete the following sentences with the words given below. Change the form if necessary.

collaborate	decommission	incarnation	watershed	underserved
unleash	objectionable	controversy	myriad	proliferation
pervasive	reshape	blackmail	flurry	meteoric

1. She agreed to _____ with him in writing her biography.
2. The officers were still reluctant to _____ their troops in pursuit of a defeated enemy.
3. The government insisted that it would not be _____ by violence.
4. A recent study of retail firms confirmed that IT has become _____ and relatively easy to acquire.
5. The proposals to reduce the strength of the army have been the subject of much _____ .
6. A _____ of diplomatic activity aimed at ending the war.
7. The meeting has the potential to be a _____ event.
8. Microsoft expects to _____ the older servers by the end of the year.
9. The secret of Walmart's _____ increase over the past five decades has been its obsession with low prices.
10. If they succeed, then they will _____ the political and economic map of the world.

Part 2　Reading

Text A

The History of Computers

The first counting device was used by the primitive people. They used sticks, stones and bones as counting tools. With human mind and technology improved, more computing devices were developed. Some of the popular computing devices starting with the first to recent ones are described below.

■**Abacus** (Figure 1 – 1)

The history of computer began with the birth of abacus which was believed to be the first computer. It was said that Chinese invented Abacus around 4,000 years ago.

It was a wooden rack which had metal rods with beads mounted on them. The beads were moved by the abacus operator according to some rules to perform arithmetic calculations. Abacus is still used in some countries like China, Russia and Japan. An image of this tool is shown on the right.

Figure 1 – 1

■**Napier's Bones** (Figure 1 – 2)

It was a manually-operated calculating device which was invented by John Napier (1550 – 1617) of Merchiston. In this calculating tool, he used 9 different **ivory** strips or bones marked with numbers to multiply and divide. So, the tool became known as "Napier's Bones." It was also the first machine to use the **decimal point**.

Figure 1 – 2

■**Pascaline** (Figure 1 – 3)

Pascaline was also known as Arithmetic Machine or Adding Machine. It was invented between 1642 and 1644 by a French mathematician-philosopher Biaise Pascal. It was believed that it was the first mechanical and automatic calculator.

Figure 1 – 3

Pascal invented this machine to help his father, a tax accountant. It could only perform addition and **subtraction**. It was a wooden box with a series of **gears** and wheels. When a wheel was **rotated** with one revolution, it rotated the neighboring wheel. A series of windows was given on the top of the wheels to read the totals. An image of this tool is shown in Figure 1 – 3.

■**Stepped Reckoner or Leibnitz wheel** (Figure 1 – 4)

It was developed by a German mathematician-philosopher Gottfried Wilhelm Leibnitz in 1673. He improved Pascal's invention to develop this machine. It was a digital mechanical calculator which was called the steppedreckoner as it was made of **fluted** drums instead of gears. See the image on the right.

Figure 1 – 4

■**Difference Engine** (Figure 1 – 5)

In the early 1820s, it was designed by Charles Babbage who was known as "Father of Modern Computer". It was a mechanical computer which could perform simple calculations. It was a steam-driven calculating machine designed to solve tables of numbers like **logarithm** tables.

Figure 1 – 5

■**Analytical Engine** (Figure 1 – 6)

This calculating machine was also developed by Charles Babbage in 1830. It was a mechanical computer that used **punch-cards** as input. It was capable of solving any mathematical problem and storing information as a permanent memory.

Figure 1 – 6

■**Tabulating Machine** (Figure 1 – 7)

It was invented in 1890, by Herman Hollerith, an American statistician. It was a mechanical tabulator based on punch cards. It could **tabulate** statistics and record or sort data or information. This machine was used in the 1890 U. S. Census. Hollerith also started the Hollerith's Tabulating Machine Company which later became International Business Machine (IBM) in 1924.

Figure 1 – 7

■**Differential Analyzer** (Figure 1 – 8)

It was the first electronic computer introduced in the United States in 1930. It was an **analog** device invented by Vannevar Bush. This machine had vacuum tubes to **switch** electrical signals to perform calculations. It could do 25 calculations in few minutes.

Figure 1 – 8

■**Mark I** (Figure 1 – 9)

The next major changes in the history of computer began in 1937 when Howard Aiken planned to develop a machine that could perform calculations involving large numbers. In 1944, Mark I computer was built as a partnership between IBM and Harvard. It was the first programmable digital computer.

Figure 1 – 9

■**Generations of Computers**

A generation of computers refers to the specific improvements in computer technology with time. In 1946, electronic pathways called circuits were developed to perform the counting. It replaced the gears and other mechanical parts used for counting in previous computing machines.

In each new generation, the circuits became smaller and more advanced than the previous generation circuits. The **miniaturization** helped increase the speed, memory and power of computers. There are five generations of computers which are described below.

First Generation Computers

The first generation (1946 – 1959) computers were slow, huge and expensive. In these computers, vacuum tubes were used as the basic components of CPU and memory. These computers were mainly depended on **batch operating system** and punch cards. Magnetic tape and paper tape were used as output and input devices in this generation.

Some of the popular first generation computers are ENIAC, EDVAC, UNIVAC-1, IBM-701 and IBM-650.

Second generation computers

The second generation (1959 – 1965) was the era of the transistor computers. These computers used transistors which were cheap, **compact** and consuming less power; it made transistor computers faster than the first generation computers.

In this generation, magnetic cores were used as the primary memory and magnetic disc and tapes were used as the secondary storage. **Assembly language** and **programming languages** like COBOL and FORTRAN, and Batch processing and multi-programming operating systems were used in these computers.

Some of the popular second generation computers are IBM 1620, IBM 7094, CDC 1604, CDC 3600 and UNIVAC 1108.

Third generation computers

The third generation computers used **integrated circuits** (ICs) instead of transistors. A single IC can pack huge number of transistors which increased the power of a computer and reduced the cost. The computers also became more reliable, efficient and smaller in size. These computers used remote processing, time-sharing, multi-programming as operating system. Also, the high-level programming languages like FORTRON-II TO IV, COBOL, PASCAL PL/1, ALGOL-68 were used in this generation.

Some of the popular third generation computers are IBM-360 series, Honeywell-6000 series, PDP, IBM-370/168 and TDC-316.

Fourth generation computers

The fourth generation (1971 – 1980) computers used very large scale integrated (VLSI) circuits—a chip containing millions of transistors and other circuit elements. This chips made this generation computers more compact, powerful, fast and affordable. These computers used real time, time sharing and distributed operating system. The programming languages like C, C++, DBASE were also used in this generation.

Some of the popular fourth generation computers are DEC 10, STAR 1000, PDP 11, CRAY-1(Super Computer) and CRAY-X-MP(Super Computer).

Fifth generation computers

In fifth generation (1980 –) computers, the VLSI technology was replaced with ULSI (Ultra Large Scale Integration). It made possible the production of microprocessor chips with ten million electronic components. These generation computers used parallel processing hardware and AI (Artificial Intelligence) software. The programming languages used in this generation were C, C++, Java, .Net, etc.

Some of the popular fifth generation computers are Desktop, Laptop, NoteBook, UltraBook and ChromeBook.

(1064 words)

【Words, Phrases and Expressions】

ivory /ˈaɪvəri/　n. 象牙,象牙制品;象牙色,乳白色

decimal /ˈdesɪməl/　adj. 十进位的,小数的　n. 小数;十进制

subtraction /səbˈtrækʃən/　n. 减去;(数)减,减法

gear /gɪr/ n. 排挡,齿轮;(特定用途的)器械,装置
　　　　v. 使变速,使调挡

rotate /ˈrəʊˌteɪt/　v. (使)旋转,(使)转动;(人员)轮换,轮值;轮种,轮作;定期调换地点
　　　　(或位置)

fluted /flutɪd/　adj. 外部有凹槽纹的

logarithm /ˈlɒgəˈrɪðəm/　n. [数]对数

tabulate /ˈtæbjəˌleɪt/　v. 制成表格

analog /ˈænəlɒg/　adj. 模拟的,类比的　n. 模拟,相似物

switch /swɪtʃ/　v. (用开关)打开,关闭;(使)改变,转变;交换,调换;转移(注意力);(与
　　　　某人)交换工作,调班

miniaturization /ˌmɪniətʃərɪˈzeɪʃən/　n. 小型化,微型化

batch /bætʃ/　n. 一批,一批生产量　v. 分批处理

compact /ˈkɒmpækt/　adj. 小型的,袖珍的;紧凑的;紧密的,坚实的;矮小而健壮的;(演
　　　　讲,作品)简洁的,简练的
　　　　n. 合约,协定;小型汽车;袖珍物
　　　　v. 压紧,压实;缩短,精简;订立(或签订)(协定)

decimal point　小数点

punch-card　穿孔卡片;打孔卡

batch operating system 批处理操作系统

assembly language　汇编语言

programming language　编程语言

integrated circuit　集成电路

 Notes

1. ENIAC

ENIAC(Electronic Numerical Integrator and Computer) was the first electronic general-purpose computer. It was Turing-complete, digital, and capable of being reprogrammed to solve a full range of computing problems.

2. EDVAC

EDVAC(Electronic Discrete Variable Automatic Computer) was one of the earliest electronic computers. Unlike its predecessor the ENIAC, it was binary rather than decimal, and was designed to be a stored-program computer.

3. UNIVAC

UNIVAC (Universal Automatic Computer) was a line of electronic digital stored-program computers starting with the products of the Eckert-Mauchly Computer Corporation.

4. CDC 1604

CDC 1604 was a 48-bit computer designed and manufactured by Seymour Cray and his team at the Control Data Corporation (CDC).

5. PDP

PDP is the abbreviation for Programmed Data Processor, a series of minicomputers made by Digital Equipment Corporation (DEC).

6. CRAY-1

Cray-1 was a supercomputer designed, manufactured and marketed by Cray Research which was the first supercomputer to successfully implement the vector processor design. These systems improved the performance of math operations by arranging memory and registers to quickly perform a single operation on a large set of data.

7. ChromeBook

A Chromebook is a laptop or tablet running the Linux-based Chrome OS as its operating system. Chromebooks are primarily used to perform a variety of tasks using the Google Chrome browser, with most applications and data residing in the cloud rather than on the machine itself.

 Exercise

A. Answer the following questions according to the text.

1. When did computers start?

2. Whom did Pascal invent Pascaline to help?

3. Who was known as "Father of Modern Computer"?

4. What was the first programmable digital computer?

5. What technology was used in fifth generation computers?

B. Choose the best answer to each of the following questions according to the text.

1. Which one is true according to computing devices?

   A) Abacus was a metal frame which had wooden rods.

   B) Napier's Bones was a mechanically-operated calculating device which was also the first machine to use the decimal point.

   C) Pascaline could perform multiplication, division, addition and subtraction.

D) Stepped reckoner was not made of gears but fluted drums.

2. About tabulating machine, which one is wrong?

A) Herman Hollerith, an American statistician invented this machine in the year 1890.

B) It was capable of tabulating statistics and recording data or information.

C) It was commonly used by American people in 1890.

D) Hollerith's Tabulating Machine Company later became International Business Machine.

3. In second generation computers, _____ were used as assembly language and programming languages.

A) C and C++   B) FORTRON-II TO IV

C) ALGOL-68   D) none of the above

4. Which is irrelevant according to the article?

A) Leibnitz wheel—Gottfried Wilhelm Leibnitz—Germany

B) Difference Engine—Vannevar Bush—USA

C) First Generation Computers—magnetic tape and paper tape—ENIAC

D) Third Generation Computers—integrated circuits—Programmed Data Processor

5. What is the main idea of the article?

A) The description of the popular computing devices.

B) The major changes in the history of computer.

C) Five generations of computers.

D) All of above.

 Text B

The Development of Internet

The Internet has become a vital part of the modern world, inseparable from daily life and routines. From simple computer networks to global interconnectivity and **instantaneous** wireless communications, the rapid and dramatic evolution of the Internet can help with understanding the changing nature of technology and communications.

■**Internet History Timeline—From ARPANET to World Wide Web**

1960s

The Internet as we know it didn't exist until much later, but Internet history started in the 1960s. In 1962, MIT computer scientist J.C.R. Licklider came up with the idea for a global computer network. He later shared his idea with colleagues at the U.S. Department of Defense Advanced Research Projects Agency (ARPA). Work by Leonard Kleinrock, Thomas Merrill and Lawrence G. Roberts on **packet-switching** theory **pioneered** the way to the world's first wide-area computer network. Roberts later went on to publish a plan for the ARPANET, an ARPA-funded computer network that became a reality in 1969. Over the following years, the ARPANET grew.

1970s

In 1973, Robert Kahn and Vinton Cerf **collaborated** to develop a protocol for linking multiple networks together. This later became the Transmission Control Protocol/Internet Protocol (TCP/IP), a technology that linked multiple networks together such that, if one network was brought down, the others did not collapse. While working at Xerox, Robert Metcalfe developed a system using cables that allowed for transfer of more data over a network. He named this system Alto Aloha, but it later became known as Ethernet. Over the next few years, Ted Nelson proposed using hypertext to organize network information, and Unix became popular for TCP/IP networks. Tom Truscott and Steve Bellovin developed a Unix-based system for transferring data over phone lines via a dial-up connection. This system became USENET.

1980s

Dave Farber of the University of Delaware revealed a project to build an inexpensive network using dial-up phone lines. In 1982, the PhoneNet system was established and was connected to ARPANET and the first commercial network, Telenet. This broadened access to the Internet and allowed for e-mail communication between multiple nations of the world. In 1981, Metcalfe's company 3Com announced Ethernet products for both computer workstations and personal computers; this allowed for the establishment of local area networks (LANs). Paul Mockapetris, Jon Postel and Craig Partridge created the Domain Name system, which used **domain** names to manage the increasing number of users on the Internet. In 1985, the first domain was registered: symbolics.com, a domain belonging to a computer manufacturer.

1990s

In 1990, ARPANET was **decommissioned.** Tim Berners-Lee and his colleagues at CERN developed hypertext markup language (HTML) and the uniform resource locator (URL), **giving birth to** the first **incarnation** of the World Wide Web. A **watershed** year for the Internet came in 1995: Microsoft launched Windows 95; Amazon, Yahoo and eBay all launched; Internet Explorer launched; and Java was created, allowing for **animation** on websites and creating a new **flurry** of Internet activity. In 1996, Congress passed the Communications Decency Act in an effort to combat the growing amount of **objectionable** material on the Internet. John Perry Barlow responded with an essay, *A Declaration of the Independence of Cyberspace*. Google was founded in 1998. In 1999, the music and video piracy **controversy** Intensified with the launch of Napster. The first Internet virus capable of copying and sending itself to a user's address book was discovered in 1999.

2000s

In 2000, people saw the rise and burst of the dotcom bubble. While **myriad** Internet-based businesses became present in everyday life, the Dow Jones industrial average also saw its biggest one-day drop in history up to that point. By 2001, most publicly traded dotcom companies had been gone. It's not all bad news, though; the 2000s saw Google's **meteoric** rise to domination of the search engine market. This decade also saw the rise and **proliferation** of Wi-Fi — wireless Internet communication — as well as mobile Internet devices like smartphones and, in 2005, the first-ever Internet cat video.

■The Future of the Internet

Why is data so important?

At the heart of the Internet is data—massive amounts of it. Volumes of data are growing at a rate of 40% per year and had increased 50 times by 2020 (Figure 1 – 10). A measure of the speed of growth is the estimate from Singapore-based Aureus Analytics that 90% of all data in the world has been created in the last two years.

As mobile usage goes up, so do personal data volumes. Over half (51%) of all Internet users worldwide are in Asia.

But the amount of industrial data is growing even faster. Today Gartner estimates there are about 4. 9 billion connected devices covering products from cars, homes, appliances and industrial equipment. This had reached 25 billion by 2020, driven by initiatives—introducing smart meters and more efficient street lighting.

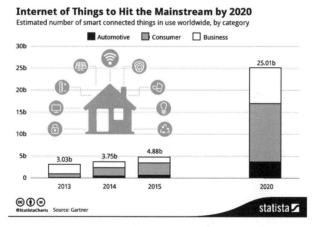

Figure 1 – 10

What opportunities and challenges will the Internet bring?

The World Economic Forum launched the Future of the Internet Initiative (FII) in 2015 to help strengthen trust and expand cooperation on Internet-related challenges and opportunities. Here are the four main themes of the FII project:

1. Transformation

The Internet is **reshaping** public and private sector structures, creating new markets for business. Another opportunity and challenge is to find new sources of income and value from digital transformation. New propositions such as city car-hire schemes are creating new partnerships, business models and platforms for bringing buyers and sellers together.

2. Access

There are still more people offline than on. The United Nations has included universal affordable Internet access in its Global Goals, which collectively aim to reduce poverty, advance health and education and improve the environment. The goals recognize that the Internet can aid global development by connecting neglected and **underserved** communities around the world.

3. Governance

As the Internet becomes more **pervasive**, so do the tensions between national interests and global inter-operability. The rules governing the cyberworld are being tested on a daily basis as governments and individuals raise concerns about data security and privacy. The innovative potential **unleashed** by digital technologies is also raising questions about rules and ethics, as well as societal benefits and costs.

4. Security

Incidents of cybercrime are rising significantly. In the United Kingdom, 2.5 million computer-related criminal offences were reported in the 12 months **leading up to** June 2015.

The year 2015 saw another significant escalation in the rise of cyber threats. The United States had the most attacks, with its industrial sector, governments and education institutions the most frequent targets.

Cybercrime is considered to be the number one motivation behind the attacks, followed by hacking and cyber **espionage**. The increase in the range and sophistication of attacks, as well as the potential brand damage due to loss of data or **blackmail**, have **pushed** cybersecurity **up** the agenda for large and small organizations.　　(1127 words)

【Words, Phrases and Expressions】

instantaneous /ˌɪnstənˈteɪniəs/　*adj.* 瞬间的,立刻的

packet-switching /ˈpækɪtˌswɪtʃɪŋ/　〔计〕分组交换

pioneer /ˌpaɪəˈnɪə/　*n.* 拓荒者;先锋　*v.* 做先锋,倡导;开辟(道路)

collaborate /kəˈlæbəreɪt/　*v.* 合作,协作;勾结,通敌

domain /dəʊˈmeɪn/　*n.* 领域,范围;领土,势力范围;(因特网上的)域

decommission /diːkəˈmɪʃn/　*v.* 正式停止使用(武器、核电站等)

incarnation /ɪnkɑːˈneɪʃn/　*n.* 化身;道成肉身;典型

watershed /ˈwɔːtəʃed/　*n.* (美)流域;分水岭;集水区;转折点

animation /ˌænɪˈmeɪʃn/　*n.* 活力,生气,热烈;动画片,动画游戏;动画制作

flurry /ˈflʌri/　*n.* 慌张;疾风;飓风;骚动;一阵忙乱(或激动、兴奋等);阵雪(或雨等);
　　　(同时出现的)一系列事物;一阵混乱

objectionable /əbˈdʒekʃənəbl/　*adj.* 讨厌的;会引起反对的;有异议的

controversy /ˈkɒntrəvɜːsi/　*n.* 争论,争议

myriad /ˈmɪriəd/　*n.* 无数,大量

meteoric /miːtiˈɒrɪk/　*adj.* 大气的;流星的;疾速的

proliferation /prəˌlɪfəˈreɪʃn/　*n.*（数量的）激增，剧增；（细胞、组织、有机体的）繁殖，增
　　　　生；大量

reshape /ˌriːˈʃeɪp/　*vt.* 改造；再成形

underserved /ˌʌndəˈsɜːrvd/　*adj.* 服务不周到的；服务水平低下的

pervasive /pəˈveɪsɪv/　*adj.* 弥漫的，遍布的

unleash /ʌnˈliːʃ/　*v.* 释放出，发泄（力量、感情等）；放开，解除对……的限制

espionage /ˈespiɑːʒ/　*n.* 间谍行为，谍报活动

blackmail /ˈblækmeɪl/　*n.* 勒索，敲诈；胁迫，威胁；勒索金　*v.* 敲诈，勒索；要挟，胁迫

give birth to　产生；造成

leading up to　在……之前

push up　增加；提高；向上推

 ## *Notes*

1. MIT

MIT is the abbreviation for Massachusetts Institute of Technology.

2. ARPA

ARPA is the abbreviation for Advanced Research Projects Agency.

3. Ethernet

Ethernet is a system for connecting a number of computer systems to form a network which is a local-area network protocol featuring a bus topology and a 10 megabit per second data transfer rate.

4. Unix

Unix is an operating system which can be used by many people at the same time.

5. CERN

CERN is the abbreviation for *Conseil Europeen pour la Recherche Nucleaire*.

6. FII

FII is the abbreviation for the Future of the Internet Initiative.

 ## *Exercise*

A. Answer the following questions according to the text.

1. When did Internet start?

2. What was the first domain of the computer manufacturer?

3. When was the watershed year of Internet? Why?

4. What is the core for Internet?

5. Why are the rules governing the online world being tested every day?

B. Read the article and decide whether the following statements are TRUE (T) or FALSE (F).

(   ) 1. In the 1990s, the first embodiment of **WWW** was created.

(   ) 2. Google was rapidly on the rise as the dominator of the search engine field in 1998.

(   ) 3. A slight majority of all Internet users worldwide are in Asia.

(   ) 4. That incidents of cybercrime are increasing slightly is the main theme of FII.

(   ) 5. The faster-growing data is the reason for putting cybersecurity on the agenda.

Part 3　Extending Skill: Note-Taking

For different types of information, different forms are needed to facilitate note-taking. The following table shows the note-taking forms you might use when taking down different information. There is also a picture showing what different note-taking forms look like.

Table 1 – 1　Different Information Type and Note-Taking Form

Information Type	Note-Taking Form
question and answer/problem and solution	headings and notes⑤
classification and definition	tree diagram②/spidergram⑦
advantage and disadvantage	two columns③
compare and contrast	table④
cause and effect	spidergram⑦
sequence of events	timeline⑥
stages of a process	flowchart①
facts and figures	table④

续表

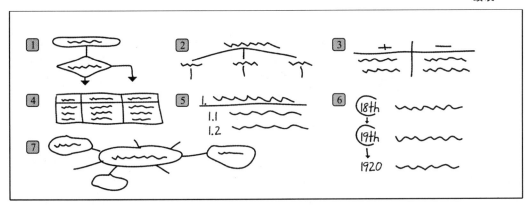

【Exercise】

Read the following lecture introduction. Choose a form to take notes.

In today's session, we are going to look at computer bugs. We will discuss four areas about it: appearing time, definition, causes and effects. The first bug appeared in 1945 which is defined as flaw in program. When it comes to causes, there are two reasons—mistakes in source code or design and different parts of program interacting in unpredictable way. As for effects, several points are mentioned. Firstly, the effects are minimal; secondly, program freezes/OS crashes; thirdly, security problems are caused, such as viruses; Lastly, the effects are serious.

Part 4　Optional Exercises

A. Translate the following paragraph into Chinese.

In primitive society, people used sticks, stones, and bones as counting tools. With the technology advanced and the human mind improved, more computing devices were developed like abacus etc. Around 4,000 years ago, the Chinese invented the abacus, and it was believed to be the first computer. The history of computers began with the birth of the abacus. Abacus was basically a wooden rack that had metal rods with beads. Beads were moved by the abacus operator according to some rules to perform arithmetic calculations. In some countries like China, Russia, and Japan, the abacus is still used by their people.

B. Translate the following paragraph into English.

　　自诞生以来,互联网发生了巨大的变化。所有迹象都表明,它将继续以难以预测的方式发生变化。因此,专业人士有必要走在潮流的前面,在未来的职业生涯中最大限度地发挥自己的潜力。了解当前和未来数字环境的最好方法之一是在课堂上更多地了解它。在线商业管理课程将帮助您启动您的职业生涯,并在我们的动态经济中预测新的技术突破。

C. Critical thinking.

　　From simple computer networks to wireless communications, Internet has gone through the rapid and dramatic evolution. Enjoying the various benefits brought by the Internet, people can not help but worry about such ethical problems as data security and privacy. In order to control the potential hazards of Internet, kinds of rules and regulations should be established by countries, companies, even individuals. What efforts will you make to minimize the potential security risks aroused by Internet?

Unit 2

Internet Application

Part 1   Vocabulary

A. Choose the explanations in Column B that best match the terms in Column A.

Column A	Column B
1. information superhighway	a. enterprises that conduct electricity in particular conditions better than insulators
2. electronic catalogue	b. a complete list of items in an electronic form
3. semiconductor industries	c. the amounts and rates of sending and receiving data via cables or wireless systems
4. Fiber-optic cable	d. sending and receiving data in an electronic way
5. volume and speed of data transmission	e. a large electronic network used for sending and receiving information
6. a two-lane highway	f. a thick wire, or a group of wires made of optical fibers that can transmit information at the speed of light
7. the electronic transmission of data	g. devices that allow a person to enter data into computer, view information, or control the operation of compute
8. online transactions and exchanges	h. man-made or virtual world
9. synthetic universe	i. a main road having a lane for traffic in each direction
10. computer terminals	j. conducting or carrying out business, giving or receiving sth. on the Internet

B. Use the words from Exercise A to label the following pictures.

1. _____

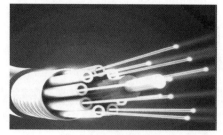

2. _____

3. _____

4. _____

5. _____

6. _____

C. Complete the following sentences with the words given below. Change the form if necessary.

catalogue	edition	feedback	transmit	format
wield	keyboard	correspondent	profit-minded	vacuum
specialize	synthetic	crucial	insure	launch

1. The campaign to wipe out illiteracy _____ out with great vigour.
2. We _____ in producing kinds of kitchen utensils and hotel tools.
3. The company has _____ all the workers who use dangerous machines.
4. The tension soon _____ itself to all the members of the crowd.
5. I got a _____ from one of the big department stores.
6. There was little negative _____ from our questionnaire.
7. The success of this experiment is _____ to the project as a whole.
8. They said foreign _____ who stayed too long in a place went blind.

9. The trade unions were afraid the government would _____ a big stick over them.

10. Then we learned to extract raw materials to create our own new _____ materials.

Part 2　Reading

 Text A

The Information Superhighway

Are you too tired to go to the video store but you want to see the movie *Beauty and the Beast* at home? Want to listen to your favorite guitar player's latest jazz **cassette**? Need some new reading material, like a magazine or book? No problem. Just sit down in front of your home computer or TV and enter what you want, when you want it, from an **electronic catalogue** containing thousands of titles.

Your school has no professors of Japanese, a language you want to learn before visiting Japan during the coming summer holiday. Don't worry. Just **sign up for** the language course offered by a school in another district or city, have the latest **edition** of the course teaching materials sent to your computer, and attend by video. If you need extra help with a translation assignment or your pronunciation, a tutor can give you **feedback** via your computer.

Welcome to the information superhighway.

While nearly everyone has heard of the information superhighway, even experts **differ on** exactly what the term means and what the future it promises will look like. Broadly speaking, however, the superhighway **refers to** the union of today's broadcasting, cable, video, telephone, and computer and **semiconductor** industries into one large all-connected industry.

Directing the union are technological advances that have made it easier to store and rapidly **transmit** information into homes and offices. **Fiber-optic** cable, for example — made up of hair-thin glass fibers — is a tremendously efficient **carrier** of information. Lasers shooting light through glass fiber can transmit 250,000 times as much data as a standard telephone wire, or tens of thousands of paragraphs such as this one every second.

The greatly increased volume and speed of data transmission that these technologies permit can be compared to the way in which a highway with many lanes allows more cars to move at faster speeds than a two-lane highway — hence, the information superhighway.

The closest thing to an information superhighway today is the Internet, the system of linked computer networks that allows up to 25 million people in 135 countries to exchange information.

But while the Internet primarily moves words, the information superhighway will soon **make routine** the electronic transmission of data in other **formats**, such as **audio files** and images. That means, for example, that a doctor in Europe who is particularly learned will be able to treat patients in America after viewing their records via computer, deciding the correct **dose** of medicine to give the patient, or perhaps even remotely controlling a **blade wielding** robot during surgery.

"Sending a segment of video mail down the hall or across the country will be easier than typing out a message on a **keyboard**," predicts one **correspondent** who **specializes in** technology.

The world is on "the eve of a new era," says the former United States Vice President Al Gore, the Clinton administration's leading high-technology advocate. Gore wants the federal government to **play the leading role** in shaping the superhighway. However, in an era of smaller budgets, the United States government is unlikely to **come up with** the money needed during the next 20 years to construct the superhighway.

That leaves private industry — computer, phone, and cable companies — to move into the **vacuum** left by the government's absence. And while these industries are pioneering the most exciting new technologies, some critics fear that profit-minded companies will only develop services for the wealthy. "If left **in the hands of** private enterprise, the data highway could become little more than a **synthetic** universe for the rich," worries Jeffrey Chester, president of the Center for Media Education in Washington, D.C.

"Poor people must also **have access to** high technology," says another expert. "Such access will be **crucial** to obtaining a high-quality education and getting a good job. So many transactions and exchanges are going to be made through this medium — banking, shopping, communication, and information — that those who have to rely on the postman to send their **correspondence** risk really falling behind," he says.

Some experts were alarmed earlier this year when **diagrams** showed that four regional phone companies who are building components of the superhighway were only connecting wealthy communities.

The companies denied they were avoiding the poor, but conceded that the wealthy would likely be the first to benefit. "We had to start building some place," says a spokesman for one of the companies, "and that was in areas where there are customers we believe will buy the service. This is a business."

Advocates for the poor want the companies building the data highway to devote a portion of their profits to **insuring** universal access. Advocates of universal access have already **launched** a number of projects of their own. In Berkeley, California, the city's Community Memory Project has placed computer **terminals** in public buildings and subway stations, where a message can be sent for 25 cents. In Santa Monica, California, computers have replaced typewriters in all public libraries, and anyone, not just librarians, can send correspondence via computer.

Many challenges face us as we move closer to the reality of the information superhighway. In order for it to be of value to most people, individuals need to become informed about what is possible and how being connected will be of benefit. The possibilities are endless but in order for the information superhighway to become a reality, some concrete steps need to be taken to get the process started.　(911 words)

【Words, Phrases and Expressions】

cassette /kəˈset/　n.(盛胶卷或磁带等的)封闭小盒;盒式录音机

electronic /ɪˌlekˈtrɒnɪk/　adj.电子(学)的;关于电子(学)的;与电子装置有关的

catalogue /ˈkætəlɒg/　n.目录,目录册;系列　v.将(某物)编入目录

edition /ɪˈdɪʃn/　n.(书、报等的)版次;版本

feedback /ˈfiːdbæk/　n.反馈信息;反馈

semiconductor /ˌsemɪkənˈdʌktə/　n.半导体

transmit /trænsˈmɪt/　v.传输;传播;传送;传递;传导

fiber-optic /ˈfaɪbəˈɒptɪk/　adj.光学纤维的

carrier /ˈkærɪə/　n.载体;病毒携带者

format /ˈfɔːmæt/　n.格式;样式;计划;安排

audio /ˈɔːdɪəʊ/　adj.声音的;音频的

file /faɪl/　n.(计算机)文件;文档;档案

dose /dəʊs/　n.(药的)一剂;一服

blade /bleɪd/　n.刀片;刀刃

wield /wiːld/　vt.挥动;使用

keyboard /ˈkiːbɔːd/　n.键盘

correspondent /ˌkɒrəˈspɒndənt/　n.记者;通讯

vacuum /ˈvækjuəm/　n.真空状态;真空

synthetic /sɪnˈθetɪk/　adj.人造的;合成的

crucial /ˈkruːʃəl/　adj.至关重要的,关键的

correspondence /ˌkɒrɪˈspɒndəns/　n.信件;通信

diagram /ˈdaɪəˌgræm/　n.图解;示意图

insure /ɪnˈʃʊə/　v.保证,担保;给……保险;投保

launch /lɑːntʃ/　vt.发射(导弹、火箭等);发起,发动,使……下水

　　　　　　　　vi.开始;下水;起飞

terminal /ˈtɜːmɪnl/　n.计算机终端;候机楼;终点站;码头

sign up for　报名参加,招收

differ on/about　在……上观点不一致

refer to　提到,指的是……,与……有关

make sth. routine　使某物普遍

specializes in　专门从事,专修

play a leading/major/key role　起主要/关键作用

come up with　拿出;提供

in the hands of sb./in sb.'s hands　由某人负责处理

have access to　有权使用(接触)

 ## *Exercise*

A. Answer the following questions according to the text.

1. What can you do if you are too tired to go to the video store but you want to see the movie *Beauty and the Beast* at home?

2. Do people share the same opinion about what the information super highway is when they have talked about it?

3. What is fiber-optic cable? What's the function of it?

4. How did the information superhighway get it's name?

5. What example is used to illustrate the information superhighway will soon make routine the electronic transmission of data in other formats?

B. Choose the best answer to each of the following questions according to the text.

1. What does the information of "superhighway" refer to?

    A) Broadcasting, cable, video, telephone.

    B) Semiconductor industries.

    C) Computer.

    D) The union of the above all into one large all-connected industry.

2. Which of the following statements is NOT true?

    A) Lasers shooting light through glass fiber can transmit 250,000 times as much data as a standard telephone wire.

    B) The closest thing to an information superhighway today is the Internet.

    C) The Internet primarily moves words, the information superhighway will soon make routine the electronic transmission of data in the same formats.

    D) If you need extra help with a translation assignment or your pronunciation, a tutor can give you feedback via your computer.

3. According to the passage, to obtain a high-quality education and get a good job, which of the following is necessary for poor people?

A) They must have access to high technology.

B) They must study hard.

C) They must have a computer.

D) They must ask for help from the wealthy.

4. Which of the following are used to move into the vacuum left by the government's absence to construct the superhighway?

A) Computer companies.   B) Phone companies.

C) Cable companies.   D) All of the above.

5. What can you infer from the passage?

A) People have to rely on the postman to send their correspondence.

B) Many challenges face us as we move closer to the reality of the information superhighway.

C) Typing out a message on a keyboard will be easier than sending a segment of video mail down the hall or across the country.

D) Advocates for the wealthy want the companies building the data highway to devote a portion of their profits to insuring universal access.

 Text B

What is Internet Application

Internet Applications can be described as the type of applications that use the Internet for operating successfully, that is, by using the Internet for fetching, sharing and displaying the information from the respective server systems. It can be accessed only with the help of the Internet facility, and it cannot be functional without the Internet. These applications can **be classified as** electronic devices based, automated digital technology, industrial Internet, smartphones based, smart home-based, smart grids, smart city, and other major applications.

■**Services of Internet Application**

Here are some of the Internet application explained in detail:

The Internet has many few major applications like electronic mail services, web browsing and peer to peer networking. The use of e-mail increases because of its several features like **attachments**, messages and data usage.

The attachment feature such as word documents, excel sheets, and **graphical** media is possible because of Multipurpose Internet Mail Extensions, but the result is that traffic volume caused by mail is **calibrated** in terms of data packets in the network.

Electronic mail services became a vital part of personal and professional communication method, saving time and cost. The data is transmitted and received securely by **encryption**. The electronic tickets for transport and sport are received in the mail.

The web browser is a critical application of the Internet and is highly commercial, which is dominated by Microsoft and highly influenced by WWW—World Wide Web.

The web browser is free and available as an **open-source** model that enriches the minds of future generations. The open-source has been developed and **deployed** on a **modular** basis since the source code is accessible only with few usage restrictions. The open-source feature has been **integrated** to file managers and web browsers.

Other important application which is potentially needed in Internet application is peer-to-peer networking. This P2P networking is a **dynamic** method that is based on the exchanging of physical resources like hard drives, files, processors and other intelligent features.

■Top Application of Internet

Here are the top eight Internet applications listed below.

1. Smart Home

Smart Home has become the **evolutionary** ladder in residential and developing as common as smartphones. It is a special feature of Google and now deployed in many areas to make life convenient and user-friendly. The smart home is designed to save time, money and energy.

2. Electronic Devices

Electronic devices like wearables are **installed with** different sensors and software, which gather data and information of the user where data is processed to give required info about the user. The devices mainly used to monitor fitness, entertainment, and health. They mostly work on **ultra-low** power and available in small sizes.

3. Automated Digital Technology

The automated digital technology has concentrated on the **optimization** of vehicles and their internal functions. The automated car is designed with special features that give a comfort zone to passengers with onboard sensors and Internet establishment. Popular companies like Tesla, Apple, BMW, Google is yet to aboard their revolution in the automobile industry by installing excellent features.

4. Industrial Internet

The industrial Internet is Investing in industrial engineering with artificial intelligence and data analytics to build brilliant machines. The important moto is to build smart machines that are accurate and **compatible** with a human. It holds vast potential with good quality and reliability. The applications are deployed for tracing the goods to be delivered, real-time data regarding retails and supplies that increase the efficiency of the business's supply chain and productivity.

5. Smart City

A smart city is another major implementation of the Internet, which is employed for smart **surveillance**, water distribution, automatic transportation, environment monitoring. People are **prone to** pollution, improper supplies and shortage of sources, and the installation of traffic sensors solves irregular traffic flow, and the app is developed to report the **municipal** systems. Citizens can able to diagnose simple **malfunctions** in meter and can report to the electricity system via electricity board applications or websites, and they can also find available **slots** for vehicle parking easily in sensor systems.

6. Smartphones

Smartphones are also used for retailers and customers to stay connected for their business transactions, even out of the store. They have using Beacon technology to help business people to provide smart service to the client. They can track the products and enhance the store dashboard and deliver **premium** order before the scheduled date, even in **congested** traffic areas.

7. Smart Grids

The idea **applied in** smart grids is to gather data in an automated way to analyze the attribute of electricity. Consumers improve the efficiency and economics of usage. Smart grids can easily detect the **power outage** and shortage quickly and fix them shortly.

8. Major Application

Another major application of the Internet is in healthcare as it is smart medical systems installed to diagnose and cure the disease at an earlier stage. Many machine learning **algorithms** are used in image processing and classification to detect the **fetus's abnormalities** before birth. The main aim applied in the medical field is to provide a healthier life for all by wearing connected devices. The gathered medical data of patients made the treatment easier, and a monitoring device is installed to track the sugar and blood pressure.

■**Conclusion**

As discussed, the Internet provides enormous application in all fields to reduce the complexity and on-time delivery with high quality of customer relationship management. Social media is **on-trend** to spread the news faster, which gets the people closer to solve the issue irrespective of time and place.   (918 words)

【Words, Phrases and Expressions】

attachment /əˈtætʃmənt/   *n.* 附件,附属物

graphical/ˈɡræfɪkl/   *adj.* 绘画的;计算机图形的;用图(或图表等)表示的

calibrate/ˈkælɪbreɪt/   *v.* 校准,标定(测量仪器等);精确测量,准确估量;调整(实验结果),调节

encryption /ɪnˈkrɪpʃn/　　*n.* 加密

open-source /ˌəʊpən ˈsɔːs/　*adj.* 开源的，开放的

deploy /dɪˈplɔɪ/　*v.* 部署，调度；利用

modular /ˈmɒdjʊlə/　*adj.* 组合式的；模块化的；模数的；有标准组件的

integrate /ˈɪntɪɡreɪt/　*v.* (使)合并，成为一体；(使)加入，融入群体　*adj.* 整合的

dynamic /daɪˈnæmɪk/　*adj.* 充满活力的，精力充沛的；动态的，动力的
　　　　　　　　　　　　n. 动力，活力；动态；动力学

evolutionary /ˌiːvəˈluːʃənəri/　*adj.* 进化论的，进化的；演变的，逐步发展的

ultra-low /ˈʌltrəˈləʊ/　*adj.* 超低的

optimization /ˌɒptɪmaɪˈzeɪʃən/　*n.* 最佳化，最优化

compatible /kəmˈpætəbl/　*adj.* 兼容的；可共存的；与……一致的

surveillance /səːrˈveɪləns/　*n.* 监视，监察

municipal /mjuːˈnɪsɪpl/　*adj.* 城市的，市政的；自治城市的，地方自治的；内政的

malfunction /ˌmælˈfʌŋkʃn/　*v.* 出现故障，运转失灵　*n.* 故障，失灵

slot /slɒt/　*n.* 时间，空档；位置，职位

premium /ˈpriːmiəm/　*adj.* 高价的，高品质的　*n.* 保险费；额外补贴，津贴；奖品，奖金

congest /kənˈdʒest/　*vt.* 使充血；充塞　*vi.* 充血；拥挤

algorithm /ˈælɡərɪðəm/　*n.* (尤指计算机)算法，运算法则

fetus /ˈfiːtəs/　*n.* 胎儿，胎

abnormality /ˌæbnɔːˈmæləti/　*n.* 异常；畸形，变态

on-trend /ˈɒnˈtrend/　*adj.* 盛行的；时髦的

be classified as　把……分类为……

install with　安装

prone to　倾向于；易于做某事

apply in　适用于；应用于

power outage　停电；断电

Exercise

A. Answer the following questions according to the text.

1. What is Internet application?

2. Why does the use of e-mail rise?

3. Why are some attachment features possible?

4. What does the automated digital technology focus on?

5. What are the benefits of smart grids? Could you list some mentioned in the article?

**B. Read the article and decide whether the following statements are TRUE (T)
or FALSE (F).**

(　　)1. Microsoft administrates the web browser, deeply affected by WWW.

(　　)2. P2P networking is a static way.

(　　)3. All electronic devices work on extremely low power and in small sizes.

(　　)4. Smart home is used for smart surveillance, water distribution, and environment
monitoring.

(　　)5. Social media can make people solve the problems more closely, regardless of
time and place.

Part 3　Extending Skill: Analyzing Complex Sentences

Sentences in academic and technical texts are often very long. For example,
Internet Applications can be described as the type of applications that
use the Internet for operating successfully, that is, by using the
Internet for fetching, sharing and displaying the information from
the respective server systems.(Text B, Unit 2)

You often don't have to understand every word, but you must identify the subject,
the verb and the object, if there is one.

For example, in the sentence above, we find:

• Subject: Internet Applications

• Verb: can be described as

• Object: the type of applications

Remember!

You can remove any prepositional phrases at this point to help you find the main
information, e.g., by using the Internet...

You can also remove any introductory phrase, e.g., that is,...

You must then find the main words which modify the subject, the verb and the
object or complement.

In the sentence above we find:

—What applications?—Internet

—What type of applications?—Using the Internet for operating successfully.

We can take the following sentences (a) to (d) as examples to show how to analyze
complex sentences by putting the parts of each sentence in the correct column in Table 2 – 1.

(a) While nearly everyone has heard of the information superhighway, even experts differ on exactly what the term means and what the future it promises will look like. (Text A, Unit 2)

(b) Some experts were alarmed earlier this year when diagrams showed that four regional phone companies who are building components of the superhighway were only connecting wealthy communities. (Text A, Unit 2)

(c) In Berkeley, California, the city's Community Memory Project has placed computer terminals in public buildings and subway stations, where a message can be sent for 25 cents. (Text A, Unit 2)

(d) Other important application which is potentially needed in Internet application is peer-to-peer networking. (Text B, Unit 2)

Table 2 - 1   Breaking a Complex Sentence into Constituent Parts

Sentence	Main S	Main V	Main O/C	Other V+S/O/C	Adv. phrases
(a)	experts	differ on		- While nearly everyone has heard of the information superhighway - what the term means and what the future it promises will look like	- even - exactly
(b)	Some experts	were alarmed		- when diagrams showed - that four regional phone companies were only connecting wealthy communities - who are building components of the superhighway	earlier this year
(c)	the city's Community Memory Project	has placed	computer terminals	where a message can be sent for 25 cents	- In Berkeley California - in public buildings and subway stations
(d)	Other important application	is	peer-to-peer networking	which is potentially needed in Internet application	potentially

【Exercise】

Study the following sentences from Text A and Text B and complete the following blocks by putting the parts of each sentence in the correct column.

1. The greatly increased volume and speed of data transmission that these

technologies permit can be compared to the way in which a highway with many lanes allows more cars to move at faster speeds than a two-lane highway — hence, the information superhighway. (Text A, Unit 2)

2. While these industries are pioneering the most exciting new technologies, some critics fear that profit-minded companies will only develop services for the wealthy. (Text A, Unit 2)

3. The web browser is a critical application of the Internet and is highly commercial, which is dominated by Microsoft and highly influenced by WWW—World Wide Web. (Text B, Unit 2)

4. A smart city is another major implementation of the Internet, which is employed for smart surveillance, water distribution, automatic transportation, environment monitoring. (Text B, Unit 2)

Sentence	Main S	Main V	Main O/C	Other V+S/O/C	Adv. phrases
1					
2					
3					
4					

Part 4 Optional Exercises

A. Translate the following paragraph into Chinese.

Internet Applications can be described as the type of applications that use the Internet for operating successfully, that is, by using the Internet for fetching, sharing and displaying the information from the respective server systems. It can be accessed only with the help of the Internet facility, and it cannot be functional without the Internet. These applications can be classified as electronic devices based, automated digital technology, industrial Internet, smartphones based, smart home-based, smart grids, smart city, and other major applications.

B. Translate the following paragraph into English.

当前,新一轮科技革命和产业变革加速演进,人工智能、大数据、物联网等新技术新应用新业态方兴未艾,互联网迎来了更加强劲的发展动能和更加广阔的发展空间。发展好、运用好、治理好互联网,让互联网更好造福人类,是国际社会的共同责任。各国应顺应时代潮流,勇担发展责任,共迎风险挑战,共同推进网络空间全球治理,努力推动构建网络空间命运共同体。

C. Write a composition on "How to Protect Privacy on the Internet", basing on the outline given below.

a. The issue of protecting privacy on the Internet aroused our attention.

b. How to protect privacy on the Internet?

c. The importance of protecting privacy on the Internet.

Unit 3

Software Development

A. Choose the the explanations in Column B that best match the terms in Column A.

Column A	Column B
1. integrated circuits	a. a method of making something easy to recognize or distinct
2. software	b. the technology of sending signals, images and messages over long distances by radio, telephone, television, satellite, etc.
3. discipline	c. a small microchip that contains a large number of electrical connections and performs the same function as a larger circuit made from separate parts
4. telecommunication	d. a branch of knowledge
5. notation	e. (computer science) written programs or procedures or rules and associated documentation pertaining to the operation of a computer system and that are stored in read/write memory
6. coding	f. a system of signs, symbols
7. specification	g. the first design of sth. from which other forms are copied or developed
8. prototype	h. a detailed description of how sth. is, or should be, designed or made
9. framework	i. the process by which a company, team, or individual devises and implements an overall plan to create a new software program
10. software development	j. the parts of a building or an object that support its weight and give it shape

B. Use the words from Exercise A to label the following pictures.

1. _____

2. _____

3. _____

4. _____

5. _____

6. _____

C. Complete the following sentences with the words given below. Change the
form if necessary.

element	utility	discipline	essence	specification
assimilate	demonstrate	locate	evolve	intangible
diversity	underlie	inherent	magnitude	internship

1. The BBC has just successfully _____ a new digital radio transmission system.

2. So when we interact with it , we have to integrate it , to _____ it into our system.

3. Most computer ads used to be loaded with technical _____ , virtually unreadable to the average consumer.

4. You've got to make sure that people work together across _____ .

5. Water supplies and other public _____ were badly affected.

6. The company has _____ into a major chemical manufacturer.

7. She has a better eye for similarities among cultures than for _____ .

8. If something _____ a feeling or situation, it is the cause or basis of it.

9. An acute observer usually sees the _____ of things at first sight.

10. Mainline feminism was arguing for the _____ beauty of the natural woman.

Part 2　Reading

 Text A

Software Engineering

Today, computer software has become the key element in the evolution of computer-based systems and products and one of the most important technologies on the world stage. Over the past 50 years, software has evolved from a specialized problem solving and information analysis tool to an industry in itself and becomes a dominant factor in the economies of the industrialized world.

Yet we still have trouble developing high-quality software on time and within budget. Why does it take so long to get software finished? Why are development costs so high? Why can't we find all errors before we give the software to our customers? Why do we spend so much time and effort maintaining existing programs? And why do we continue to have difficulty in measuring progress as software being developed and maintained? These questions and many others demonstrate the industry's concern about software and the manner in which it is developed—a concern that has lead to the adoption of software engineering practice.

Then, what is software? And which characteristics of software make it different from other things that human beings build? Software is a logical rather than a physical system element, and some characteristics make software special. First, it is difficult for a customer to **specify** requirements completely, and difficult for the supplier to understand fully the customer needs as well. In defining and understanding requirements, especially changing requirements, large quantities of information need to be communicated and **assimilated** continuously. Second, software is seemingly easy to change and is primarily **intangible**, and much of the process of creating software is also intangible, involving experience, thought and imagination. In addition, it is difficult to test software **exhaustively**.

Software—programs, data, and documents—**addresses** a wide range of technology and application areas, yet all software evolves according to a set of **laws** that have remained the same for over 30 years. The intent of software engineering is to provide a **framework** and a solution for building higher quality software. It includes greater emphasis on systematic development, a concentration on finding out the user's requirements, formal/semi formal specifications of the requirements of a system, demonstration of early version of a system (**prototype**), greater emphasis on trying to ensure error free code and so on.

From the time in 1968 when the phrase "software engineering" was first used at a NATO conference until the present day, software has come a long way. But it still has a very long way to go if it is to be considered as mature as other engineering **disciplines**.

Virtually all countries now depend on complex computer-based systems. National infrastructures and **utilities** rely on computer-based systems and most electrical products include a computer and controlling software. Industrial manufacturing and distribution is completely computerized, as is the financial system. Therefore, producing and maintaining software **cost-effectively** is essential for the functioning of national and international economies.

Software engineering is an engineering discipline whose focus is the cost effective development of high quality software systems. Software is abstract and intangible. It is not constrained by materials or governed by physical laws or by manufacturing processes. In some ways, this simplifies software engineering as there are no physical limitations on the potential of software. However, this lack of natural constraints means that software can easily become extremely complex and hence very difficult to understand.

The notion of software engineering was first proposed in 1968 at a conference held to discuss what was then called the "software crisis". This software crisis resulted directly from the introduction of new computer hardware based on integrated circuits. Their power made **hitherto** unrealizable computer applications a feasible proposition. The resulting software was orders of **magnitude** larger and more complex than previous software systems.

Early experience in building these systems showed that informal software development was not good enough. Major projects were sometimes years late. The software cost much more than predicted, was unreliable, was difficult to maintain and performed poorly. Software development was in crisis. Hardware costs were **tumbling** whilst software costs were rising rapidly. New techniques and methods were needed to control the complexity inherent in large software systems.

These techniques have become part of software engineering and are now widely used. However, as our ability to produce software has increased, so has the complexity of the software systems that we need. New technologies resulting from the convergence of computers and communication systems and complex graphical user interfaces place new demands on software engineers. As many companies still do not apply software engineering techniques effectively, too many projects still produce software that is unreliable, delivered late and over budget.

We have made tremendous progress since 1968 and that the development of software engineering has markedly improved our software. We have a much better understanding of the activities involved in software development. We have developed effective methods of software **specification**, design and **implementation**. New **notations** and tools reduce the effort required to produce large and complex systems.

We know now that there is no single "ideal approach" to software engineering. The wide **diversity** of different types of systems and organizations that use these systems means that we need a diversity of approaches to software development. However, fundamental notions of process and system organization **underlie** all of these techniques, and these are the **essence** of software engineering.

Software engineers can be rightly proud of their achievements. Without complex software we would not have explored space, would not have the Internet and modern telecommunications, and all forms of travel would be more dangerous and expensive. Software engineering has contributed a great deal, and as the discipline matures, its contributions in the 21st century will be even greater.   (932 words)

【Words, Phrases and Expressions】

specify /ˈspesɪfaɪ/   *vt.* 具体说明；明确规定

assimilate /əˈsəməleɪt/   *v.* 吸收

intangible /ɪnˈtændʒəbl/   *adj.* 无形的

exhaustively /ɪgˈzɔːstɪvli/   *adv.* 详尽地，彻底地

address /əˈdres/   *v.* 处理

law /lɔː/   *n.* 规则，法则

framework /ˈfreɪmwəːk/   *n.* 构架，体系

prototype /ˈprəʊtətaɪp/   *n.* 原型；雏形

discipline /ˈdɪsəplɪn/   *n.* 学科

utility /juːˈtɪləti/   *n.* 公用事业，公共设备

cost-effective /ˈkɒstɪˈfektɪv/   *adj.* 有成本效益的，划算的

hitherto /hɪðəˈtuː/   *adv.* 迄今，至今

magnitude/ˈmægnɪtjuːd/   *n.* 巨大；震级

tumble /ˈtʌmbl/　v. 使倒下,搅乱

specification /ˌspesɪfɪˈkeɪʃn/　n. 明细单;说明书

implementation /ˌɪmplɪmenˈteɪʃn/　n. 实施,执行

notation /nəʊˈteɪʃn/　n. 符号

diversity /daɪˈvɜːsəti/　n. 差异,多样性

underlie /ˌʌndəˈlaɪ/　v. 构成……的基础

essence /ˈesəns/　n. 本质,实质

 Exercise

A. Answer the following questions according to the text.

1. What role is software playing in the economies of the modern world?

2. Why do we say software is a logical system rather than a physical one?

3. What is software engineering mainly dealing with?

4. Is software influenced by physical force?

5. Why do soft engineers take pride in themselves?

B. Choose the best answer to each of the following questions according to the text.

1. Which is correct about the development of software ?

   A) It emerged with software engineering at the same time.

   B) For a half-century development, it has almost solved problems of high-quality, on-time and within-budget.

   C) It was just a specialized problem solving and information analysis tool in its early years of development.

   D) The laws which software evolves according to have changed absolutely during its development.

2. Where and when was the phrase "software engineering" first used?

   A) In a thesis in 1966.　　　　　　B) In a conference in 1968.

   C) In a journal in 1868.　　　　　　D) In a magazine in 21 century.

3. Which of the following descriptions is NOT the characteristic of software?

   A) Abstract and intangible.

   B) Not constrained by materials.

   C) Not governed by physical laws or by manufacturing processes.

   D) Easy to understand and simple to produce as there are no physical limitations.

4. What problem(s) existed widely in informal software development in the early years?

   A) Over schedule.　　　　　　　　B) Cost much more than budget.

C) Difficult to maintain.   D) All of the above.

5. Which of the following statements is wrong about the techniques in software engineering?

   A) Techniques are needed to control the complexity of the large software systems.

   B) Techniques are the essence of software engineering.

   C) Techniques are now widely used in software engineering.

   D) New technologies bring new challenges to software engineers continually.

 Text B

Software Development

Software development refers to a set of computer science activities dedicated to the process of creating, designing, deploying and supporting software. Software itself is the set of instructions or programs that tell a computer what to do. It is independent of hardware and makes computers programmable. There are three basic types:

① System software provides core functions such as operating systems, disk management, utilities, hardware management and other operational necessities.

②Programming software gives programmers tools such as text editors, **compilers**, linkers, **debuggers** and other tools to create code.

③Application software (applications or apps) helps users perform tasks. Office productivity **suites**, data management software, media players and security programs are examples. Applications also refers to web and mobile applications like those used to shop on Amazon.com, socialize with Facebook or post pictures to Instagram.

A possible fourth type is **embedded** software. Embedded software is a piece of software that is embedded in hardware or non-PC devices. It is written specifically for the particular hardware that it runs on and usually has processing and memory constraints because of the device's limited computing capabilities. Examples of embedded software include those found in dedicated GPS devices, factory robots, some calculators and even modern smartwatches. Embedded systems software is used to control machines and devices not typically considered computers — telecommunications networks, cars, industrial robots and more. These devices, and their software, can be connected as part of the Internet of Things (IoT).

As the world became more and more dependent on technology with each passing day, software **automatically** became an important organ for development. Since software is needed almost everywhere today, its development is a highly intelligent and precise process, involving various steps. Known as software development life cycle(Figure 3 – 1), these steps include planning, analysis, design, development&implementation, testing and maintenance. These steps go on to create the perfect software for clients.

Let's study each of these steps to know how the perfect software is developed.

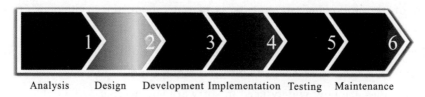

Analysis Design Development Implementation Testing Maintenance

Figure 3 – 1 The Life Cycle of Software Development

1. Analysis

Detailed analysis of the software is performed to identify **overall** requirements of client. This step is necessary to make adjustments and to ensure that software functions properly at the end.

2. Design

Once the analysis is complete, the step of designing takes over, which is basically building the **architecture** of the project. This step helps remove possible flaws by setting a standard and attempting to stick to it. The design phase models how a software application will work. Some aspects of the design include:

- Architecture. Specifies the programming language, industry practices, general design, and the use of any templates or models.
- User Interface. Defines the ways that clients interact with the software and how the software responds to input.
- Platforms. Define the platforms the software will run on, such as Apple, Android, Windows version, Linux, or even game **consoles**.
- Programming. Not only the programming language, but also the methods for solving problems and performing tasks in the application.
- Security. Defines the measures taken to protect the application and may include SSL traffic encryption, password protection, and secure storage of user **credentials**.

3. Development

Expert software developers start the actual software development process. The required components and functionalities of software are created in this very stage.

4. Testing

The testing stage assesses the software for errors and documents bugs if there are any.

5. Implementation

This stage is crucial for running the developed software assessed by all the **stakeholders**. This is to ensure the complete level of customer satisfaction.

6. Maintenance

Once the software passes through all the stages without any issues, a maintenance process is followed wherein it will be maintained and upgraded from time to time to adapt to changes. Almost every software development company follows all the six steps, leading to the reputation that the country enjoys in the software market today.

In order to build quality software, relevant information is collected from the client. This is required to to know the overall purpose of software and target audience. Project managers and stakeholders follow efficient communication strategy to hold daily and periodic meetings with clients for data gathering. Once collected, the requirements are analyzed and decide how the software will perform. Software development experts also check **feasibility** of development of software and understand expectations of clients based on gathered data.

The actual software development process is the main **crux** of the whole software development life cycle. The stage involves tasks of developing a software based on data gathered, analysis done, and design prepared in earlier stages.

An efficient team of software developers perform the detailed process of coding and add required functionalities. The overall process of developing a software is not confined to one particular industry only. The complete 6 stages of software development process is same for a varied range of industries like the following: Medical & Healthcare, Education, Banking & Finance, Media & Entertainment, Retail & Wholesale, Consumer Products, Travel, Real Estate, Automotive.

Another crucial phase of software development life cycle is testing. It helps to find bugs and any kind of error in the software. The testing stage is basically helpful to ensure the quality assurance factor of the software developed for the client. The software development service teams perform process & product **audits** along with testing during development and before delivery. Software developers also perform unit testing to make sure your product is not exposed to bugs.

Finally, after all the testing phases, final implementation of software is done by expert professionals. The software is run to see whether all the functions work properly or not. Successful implementation of project is ideally carried out by software development company that features a list of reliable technology partners. Maintenance of software created for client is required to fix issues, set major updates, and improve functionality with changing times. It is for the better use of software for clients and ensure system running smoothly.

The software development life cycle process is a thorough method to control and manage the software on a high level. The set procedure is appropriately followed no matter, whichever technology is used for the project that may include the following: PHP, Magento, Shopify, Drupal, Mean Stack, Python, WordPress, .Net, SharePoint, Cloud Solutions.   (1033 words)

【Words, Phrases and Expressions】

compiler /kəmˈpaɪlə/   n.编译程序；编纂者

debugger /diːˈbʌɡə/   n.排错程序；调试程序

suite /swiːt/   n.套件；套装

embedded /ɪmˈbedɪd/   adj.嵌入的；内嵌的

automatically /ˌɔːtəˈmætɪkli/   adv.自动地

overall /ˌəʊvərˈɔːl/   adj.总体的；全面考虑的

architecture /ˈɑːkɪˌtektʃə/   n.构造设计；体系

console /kənˈsəʊl/   n.仪表盘；操控台；（游戏）平台；落地式支座

credential /krəˈdenʃl/   n.（某人）可信任的证明

stakeholder /ˈsteɪkˌhəʊldə/   n.网络利益相关者；利益关系人

feasibility /ˌfiːzəˈbɪlɪti/   n.可行性；现实性

crux /krʌks/   n.关键；难题

audit /ˈɔːdɪt/   n.审计；审核；稽核

Exercise

A. Answer the following questions according to the text.

1. For what reason did software automatically become an important organ for development?

2. What do a good team of software often do?

3. Is testing a crucial phase of software development life cycle ? Why or why not?

4. What do software development experts do after gathering data from clients?

5. Why is maintenance of software required to fix issues, and improve functionality with changing times?

B. Read the article and decide whether the following statements are TRUE (T) or FALSE (F).

(   )1. Software refers to a set of computer science activities dedicated to the process of creating, designing, deploying and supporting software.

(   )2. Debugger is a program that helps in locating and correcting programming errors.

(   )3. Tencent meetings is an application software.

(   )4. Programming software gives programmers tools such as text editors, compilers, linkers, debuggers and other tools to design software.

(   )5. Implementation is crucial for running the developed software assessed by all the stakeholders. This is to satisfy all software development experts.

Part 3   Extending Skill: Cornell Note-Taking System

Why do you take notes? What do you hope to get from your notes? What are Cornell Notes and how do you use the Cornell note-taking system?

The Cornell Note-Taking System was originally developed by Cornell education professor, Walter Pauk. He outlined this effective note-taking method in his book *How to Study in College*. It involves the following **Five Rs:**

Record

Record in the note-taking area as many meaningful ideas and facts as possible during a lecture or discussion. Write legibly.

Reduce

As soon after as possible, summarize these facts and ideas concisely in the Cue Column. Summarizing clarifies meanings and relationships, reinforces continuity, and strengthens memory.

Recite

Cover the Note-Taking Area, using only your jottings in the Cue Column, say over the facts and ideas of the lecture as fully as you can, not mechanically, but in your own words. Then, verify what you have said.

Reflect

Draw out opinions from your notes and use them as a starting point for your own reflections on the course and how it relates to your other courses. Reflection will help prevent ideas from being inert and soon forgotten.

Review

Spend 10 minutes every week in quick review of your notes and you will retain most of what you have learned.

Here's what your paper will look like:

Header		
Lesson Title		Date
Cue or Question Column	**Note Taking Area**	
Summary Area		

Note-Taking Area: Make sure to leave large spaces in your notes to add information with the help of symbols and abbreviations.

Summary Area: Write a brief summary of that day's notes. You can choose to either write it in paragraph form or to use a graphic organizer.

Cue or Question Column: Write questions in the margins (see inside) or main ideas.

Look at this example of the Cornell Note-Taking System:

Taking Lecture Notes	03/21/23
Four steps	Taking good lecture notes involves 　• preparing for the lecture in advance 　• taking effective notes during the lecture 　• revising the notes immediately after class 　• studying the notes—as preparation for the next lecture, a test
Four aspects to prepare for a lecture	Preparation involves physical, intellectual, emotional, and spiritual preparation.
Three components of physical preparation	Physical preparation includes getting sufficient sleep to be able to remain alert in class, getting the exercise necessary to remain physically fit, and eating nutritiously.
Six components of intellectual preparation	Intellectual preparation involves 　• reading the syllabus 　• knowing about the topic(s) 　• looking ahead 　• reading assignments for possible discussions 　• reviewing previous lectures 　• conducting one's own research 　　→anticipate where the lecture will go 　　→predict the ending

Summary
There are four parts to taking good lecture notes. Preparation for a lecture should involve physical, intellectual, emotional and social preparation.

【Exercise】

Apply the Cornell note-taking system after reading the passage below. Scan the QR code on the right side to get the supplemental material about note-taking symbols and abbreviations.

What is Involved in Computer Software Development?

The software development process requires careful analysis, specification, architecture design, and implementation. Next steps include software testing, documentation, training, and ongoing user support. Several different types of computer software development models are available to help software developers create different computer programs. This development life cycle involves all steps from the initial software idea or concept to the implementation of the final product. Some of the processes used in computer software development are waterfall, iterative, incremental, and agile process models.

As the needs for developing computer software change, so will the methods of developing that software. The key ingredients in developing a software program are determined by the different type of development process used. Most software developers will use a combination or parts of each process in computer software development.

The first process in computer software development is called the waterfall model. This development model begins with taking a close look at all the requirements of potential application software, designing and integrating the actual software, conducting the necessary testing or validation, final installation and providing ongoing maintenance for the software. In this computer software development process, each phase must be completed before the next phase is started. Reviews and changes may be used after each development phase is completed.

The process of iterative and incremental computer software development lies at the other extreme of software development processes from the waterfall model. This process is used by developers when a customer may not know exactly what they need from a specific computer program. Needs analysis, programming, and component development are repeated in short cycles, or iterations, until the final computer software program is completed. This process utilizes information gathering and work on a number of smaller components to help bring full functionality to the finished program.

Agile computer software development is less structured than the waterfall or iterative/incremental development models. Developing software with the agile development model requires more creativity than structure. In this model, software is in constant flux, follows no logical process, and remains incomplete. The people developing the software and the ways in which they work together are more important than the actual process. The power of this creative software development process lies in the empowerment and collaboration of the development team as well as in it's capacity to respond to changes needed.

Part 4   Optional Exercises

A. Translate the following paragraph into Chinese.

Location-based services（LBS）are a general class of computer program-level services used to include specific controls for location and time data as control features in computer programs. Such LBS is an information service and has a number of uses in social networking today as an entertainment service, which is accessible with mobile devices through the mobile network and which uses information on the geographical position of the mobile device. This has become more and more important with the expansion of the smartphone and tablet markets as well.

B. Translate the following paragraph into English.

需求阶段有三个基本活动。第一个是问题分析或需求分析。这个活动的目的是要理解问题的要求及适应(fit into)客户组织等。第二个活动是制作软件需求规格说明书(the software requirements specification)。第三个活动是需求确认(validation)。进行需求确认活动是为了确保说明书中的需求正是客户想要的。

C. Discussion.

Agile methods are being widely accepted in the software world recently. However, this method may not always be suitable for all products. Research the advantages and disadvantage of the Agile model, discuss what you have found out with your partner based on the notes you have taken.

Unit 4

Computer Viruses and Security

Part 1　Vocabulary

A. Choose the explanations in Column B that best match the terms in Column A.

Column A	Column B
1. System Restore	a. To start up a computer system by providing it with the required electrical power and loading the startup services until the operating system is loaded
2. malicious software	b. A recovery tool for Windows that allows you to reverse certain kinds of changes made to the operating system
3. anti-virus program	c. A program that is created to search, detect, prevent and remove software viruses from your system that can harm your system
4. Internet Explorer	d. A World Wide Web browser that comes bundled with the Microsoft Windows operating system
5. boot up	e. The place where the computer holds current programs and data that are in use
6. system memory	f. Any type of software that is intended to harm or hack the user
7. disk drive	g. Security measures at the application level that aim to prevent data or code within the app from being stolen or hijacked
8. data breach	h. A type of software application used for composing, editing, formatting and printing documents
9. word processor	i. A security incident in which information is accessed without authorization
10. Application Security	j. A piece of equipment in a computer system that is used to get information from a disk or to store information on it

B. Use the words from Exercise A to label the following pictures.

1. _____

2. _____

3. _____

4. _____

5. _____

6. _____

C. Complete the following sentences with the words given below. Change the form if necessary.

modification	breach	patch	confidentiality	availability
replicate	infect	malware	configuration	activate
attachment	password	retrievable	deletion	integrity

1. This is a property of an information system whereby its _____, reliability, completeness and promptness are assured.

2. Each of the _____ programs can execute whatever instructions the virus author intended.

3. Data encryption ensures the privacy and _____ of e-mail messages.

4. Please notify Kelvin immediately of any unauthorized use of your account or any other _____ of security.

5. The references and diagrams were _____ to the document.

6. Considerable _____ of the existing system is needed in order to improve efficiency.

7. The user login and password information has already been setup and _____.

8. If this is the initial _____, the server should not be running.

9. The data is not lost when the power is turned off and easy to _____ when needed for processing.

10. Data _____ moves copies of data from one location to another location.

Part 2　Reading

 Text A

Introduction to Computer Security

The Internet has transformed our lives in many good ways. Unfortunately, this vast network and its associated technologies also have **brought in their wake**, the increasing number of security threats. The most effective way to protect yourself from these threats and attacks is to be aware of standard **cyber** security practices.

■ **What is Computer Security?**

Computer security basically is the protection of computer systems and information from harm, theft, and **unauthorized use**. It is the process of preventing and detecting unauthorized use of your computer system.

There are various types of computer security which is widely used to protect the valuable information of an organization.

■ **What are Their Types?**

One way to **ascertain** the similarities and differences among Computer Security is by asking what is being secured. For example,

- *Information Security* is securing information from unauthorized access, **modification & deletion**.

- *Application Security* is securing an application by building security features to prevent from Cyber Threats such as SQL injection, DoS attacks, data **breaches** and etc.

- *Computer Security* means securing a standalone machine by keeping it updated and **patched**.

- *Network Security* is by securing both the software and hardware technologies.
- *Cybersecurity* is defined as protecting computer systems, which communicate over the computer networks.

It's important to understand the distinction between these words, though there isn't necessarily a clear consensus on the meanings and the degree to which they **overlap** or are **interchangeable**.

So, Computer security can be defined as controls that are put in place to provide **confidentiality**, **integrity**, and **availability** for all components of computer systems. Let's elaborate the definition.

■Components of Computer System

The components of a computer system that needs to be protected are:

- *Hardware*, the physical part of the computer, like the system memory and disk drive;
- *Firmware*, permanent software that is etched into a hardware device's **nonvolatile** memory and is mostly invisible to the user;
- *Software*, the programming that offers services, like operating system, word processor, Internet browser to the user.

■The CIA Triad

Computer security is mainly concerned with three main areas (Figure 4 – 1):

Figure 4 – 1 Computer Security

- *Confidentiality* is ensuring that information is available only to the intended audience.
- *Integrity* is protecting information from being modified by unauthorized parties.
- *Availability* ensures that authorized users have timely, reliable access to resources when they are needed.

In simple language, computer security is making sure information and computer components are usable but still protected from people or software that shouldn't access it or modify it.

■Computer Security Threats

Computer security threats are possible dangers that can possibly **hamper** the normal functioning of your computer. In the present age, cyber threats are constantly increasing as the world is going digital. The most harmful types of computer security are:

Viruses

A computer virus is a **malicious** program which is **loaded into** the user's computer **without user's knowledge**. It **replicates** itself and **infects** the files and programs on the user's PC. The ultimate goal of a virus is to ensure that the victim's computer will never be able to operate properly or even at all.

Computer Worm

A computer worm is a software program that can copy itself from one computer to another, without **human interaction**. The potential risk here is that it will use up your computer **hard disk** space because a worm can replicate in great volume and with great speed.

Phishing

Disguising as a trustworthy person or business, phishers attempt to steal sensitive financial or personal information through **fraudulent** e-mail or instant messages. Phishing is unfortunately very easy to execute. You are **deluded into** thinking it's the legitimate mail and you may enter your personal information.

Botnet

A botnet is a group of computers connected in a coordinated fashion for malicious purposes. Each computer in a botnet is called a **bot**. These bots form a network of **compromised** computers, which is controlled by a third party and used to transmit **malware** or spam, or to launch attacks. A botnet may also be known as a **zombie** army.

Rootkit

A rootkit is a computer program designed to provide continued privileged access to a computer while actively hiding its presence. Once a rootkit has been installed, the controller of the rootkit will be able to remotely execute files and change system **configurations** on the host machine.

Keylogger

Also known as a keystroke logger, keyloggers can track the real-time activity of a user on his computer. It keeps a record of all the keystrokes made by user keyboard. Keylogger is also a powerful threat to steal people's login credential such as username and password.

These are perhaps the most common security threats that you'll come across. Apart from these, there are others like **spyware, wabbits, scareware, bluesnarfing** and many more. Fortunately, there are ways to protect yourself against these attacks.

■Why is Computer Security Important?

In this digital era, we all want to keep our computers and our personal **information secure** and hence computer security is important to keep our personal information protected. It is also important to maintain our computer security and its overall health by preventing viruses and malware which would impact on the **system performance.**

■Computer Security Practices

Computer security threats are becoming relentlessly **inventive** these days. There is much need for one to arm oneself with information and resources to safeguard against these complex and growing computer security threats and stay safe online. Some preventive steps you can take include:

- Secure your computer physically by:
    - Installing reliable, reputable security and **anti-virus software**;
    - **Activating** your firewall, because a firewall acts as a security guard between the Internet and your **local area network**.
- Stay up-to-date on the latest software and news surrounding your devices and perform software updates as soon as they become available.
- Avoid clicking on e-mail attachments unless you know the source.
- Change **passwords** regularly, using a unique combination of numbers, letters and **case types**.
- Use the Internet with caution and ignore **pop-ups**, drive-by downloads while surfing.
- Take the time to research the basic aspects of computer security and educate yourself on evolving cyber-threats.
- Perform daily full system scans and create a periodic system backup schedule to ensure your data is **retrievable** should something happen to your computer.

Apart from these, there are many ways you can protect your computer system. Aspects such as **encryption** and **computer cleaners** can assist in protecting your computers and its files.

Unfortunately, the number of cyber threats is increasing at a rapid pace and more sophisticated attacks are emerging. So, having a good foundation in cybersecurity concepts will allow you to protect your computer against ever-evolving cyber threats.

(1090 words)

【Words, Phrases and Expressions】

cyber /ˈsaɪbə/　*adj.*（与）网络（有关）的

ascertain /ˌæsəˈteɪn/　*v.* 查明,确定

modification /ˌmɒdɪfɪˈkeɪʃn/　*n.* 修改的行为（过程）；修改,更改；

deletion /dɪˈliːʃn/　*n.* 删除,删减

breach /briːtʃ/　*n.* 违反,破坏；(关系)中断,终止；缺口　*v.* 违反,破坏；

patch /pætʃ/　*n.* 补丁；(软件的)补丁程序

　　　　　　vt. 临时接入(通信系统)；(用补丁对程序进行)修正,打补丁

overlap /ˌəʊvəˈlæp/　*v.* （与……）互搭,（与……）重叠　*n.* (物体的)重叠部分

interchangeable /ˌɪntəˈtʃeɪndʒəb(ə)l/　*adj.* 可互换的；可交换的；可交替的

confidentiality /ˌkɒnfɪˌdenʃɪˈæləti/　*n.* 保密性,机密性

integrity /ɪnˈtegrəti/　*n.* 正直,诚实；完整性

availability /əˌveɪləˈbɪləti/　*n.* 可用性,可得性

firmware /ˈfɜːmweə/　*n.* 固件

nonvolatile /nɒnˈvɒlətaɪl/　*adj.* 非易失性的；非挥发性的

hamper /ˈhæmpə/　*v.* 阻碍,妨碍

malicious /məˈlɪʃəs/　adj. 恶意的，恶毒的

replicate /ˈreplɪkeɪt/　v. 重复，复制；自我复制

infect /ɪnˈfekt/　v. 传染，感染；污染

phishing /ˈfɪʃɪŋ/　n. 网络仿冒，网络钓鱼

fraudulent /ˈfrɔːdʒələnt/　adj. 欺诈的，诈骗的

botnet /ˈbɒtnet/　n. 僵尸网络

bot /bɒt/　n. 机器人程序；机器人

compromise /ˈkɒmprəmaɪz/　n. 折中，妥协　v. 妥协，让步

malware /ˈmælweə/　n. 恶意软件

zombie /ˈzɒmbi/　n. 僵尸；无生气的人，麻木迟钝的人

configuration /kənˌfɪɡəˈreɪʃn/　n. 布局，构造；配置

keylogger /ˈkiːlɒɡə/　n. 键盘记录器；键盘记录木马程式

spyware /ˈspaɪweə/　n. 间谍软件；间谍程序

wabbit /ˈwæbɪt/　adj. 疲倦的；筋疲力尽的

scareware /ˈskeəweə(r)/　n. 恐吓性软件

bluesnarfing /ˈbluːˌsnɑːfɪŋ/　n. 蓝牙漏洞攻击，蓝牙窃用

inventive /ɪnˈventɪv/　adj. 有创造力的，善于创新的

activate /ˈæktɪveɪt/　v. 激活，使活化

password /ˈpɑːswɜːd/　n. 密码；口令

pop-ups /pɒpˌʌps/　n. 弹出窗口；弹出式广告视窗

retrievable /rɪˈtriːvəbl/　adj. 可取回的；可补偿的；可恢复的

computer security　计算机安全

bring in one's wake　造成随之而来的

unauthorized use　未授权使用

load into　存入，输入

without one's knowledge　未告知某人

delude into　欺骗……使相信

human interaction　人机交互

hard disk　硬盘

information secure　信息安全

system performance　系统性能；系统业绩

anti-virus software　杀毒软件

local area network　局域网

case type　大小写

computer cleaner　电脑垃圾清理软件

Notes

1. SQL injection

SQL injection, also known as SQLI, is a common attack vector that uses malicious SQL code for backend database manipulation to access information that was not intended to be displayed. This information may include any number of items, including sensitive company data, user lists or private customer details.

The impact SQL injection can have on a business is far-reaching. A successful attack may result in the unauthorized viewing of user lists, the deletion of entire tables and, in certain cases, the attacker gaining administrative rights to a database, all of which are highly detrimental to a business.

2. DoS attacks

A DoS attack can disrupts or completely denies service to legitimate users, networks, systems, or other resources.

3. network security

Network security is a set of technologies that protects the usability and integrity of a company's infrastructure by preventing the entry or proliferation within a network of a wide variety of potential threats. It is worried about what is going on within the castle walls.

4. cybersecurity

Cybersecurity is a process that enables organizations to protect their applications, data, programs, networks, and systems from cyberattacks and unauthorized access. It is much more concerned with threats from outside the castle.

5. wabbits

Wabbit is one of the first instances of malware ever to exist. While simple, it can devastate a system by squandering all operating system resources until the infected system crashes.

6. drive-by downloads

Drive by download attacks specifically refer to malicious programs that install to your devices — without your consent. This also includes unintentional downloads of any files or bundled software onto a computer device.

 Exercise

A. Answer the following questions according to the text.

1. According to the passage, how many types of computer security do we have and what are the differences among them?

2. What does the CIA Triad refer to and why is it important?

3. How do phishing attacks work?

4. What is a rootkit? Is it easy to remove them?

5. Besides the preventive steps mentioned in the passage, are there any other ways to protect your computer against cyber threats?

B. Choose the best answer to each of the following questions according to the text.

1. Which of the following sentences is wrong?

   A) The basic aim of computer security is to prevent unauthorized access to computer systems and information.

   B) Cybersecurity is to protect a standalone machine by staying up-to-date and keeping it patched.

   C) Firmware is embedded into hardware devices and mostly can't be seen.

   D) Confidentiality is to control access to data and to prevent unauthorized disclosure.

2. What does the word "**compromise**" mean in the sentence "These bots form a network of **compromised** computers, which is controlled by a third party and..."?

   A) to reach an agreement

   B) to do something which is against your principles

   C) to behave in a way that is not honest

   D) expose or make liable to danger, suspicion, or disrepute

3. Which security threat consumes large volumes of memory according to the passage?

   A) Computer worm.   B) Botnet.

   C) Keylogger.   D) Wabbits.

4. What can be inferred from the passage?

   A) The number of cyber threats are increasing rapidly.

   B) If you want to prevent your computer from cyber threats, you should have a good knowledge of cybersecurity concepts.

   C) In order to safeguard against cyber threats, you had better not use free Wi-Fi.

   D) Perform daily full system scans and create a daily system backup schedule to ensure your data is retrievable.

5. Which of the following best summarizes the main idea of the passage?

   A) Computer security is very important to prevent from DoS attacks.

   B) Confidentiality, integrity and availability are the three important components of computer systems.

   C) Cyber threats have adverse influence on the normal functioning of computer system.

   D) Computer security is of great importance to protect computer systems, so preventive steps should be taken to prevent from various computer security threats.

Text B

The Symptoms and Effects of Computer Viruses

■ **What is a Computer Virus?**

A computer virus is a self-spreading piece of software. The viruses disrupt the normal functioning of a computer, damaging its software or stealing its data. The first computer virus was created in 1971 to test whether computer software could self-replicate. Interestingly, its design was inspired by nature itself. To **put** the dangers of computer viruses **into perspective**, let's consider their biological predecessors.

As you may know, abiological virus is a tiny **parasite** relying on living cells for survival. Viruses live at the expense of the host, which can be an animal, plant, or bacteria. As soon as it has infected a host cell, a virus starts reproducing, invading other cells, and spreading disease.

In the same vein, a computer virus shows up on your Mac uninvited and starts reproducing itself. If left unchecked, it can cause a lot of damage.

Unable to function without a host, this type of malware travels from computer to computer, program to program, file to file. The dependence on a host and its self-replication capabilities make a computer virus different from other types of malware.

■ **Four Stages of Virus Infection**

While some computer virus symptoms are instantly recognizable, others can go unnoticed for a long time. This is often the case as viruses do not announce their presence on infected devices right away. Usually, they go through the following four stages of infection:

1. **Dormancy**

When a virus first infects a computer, it may remain "asleep" for a while to avoid detection. Some viruses "wake up" after an infected program has been opened a certain number of times, while other viruses wait until a predetermined date. Either way, it's hard to detect the effects of a computer virus during the dormancy period.

2. Replication

After a virus "wakes up", it starts reproducing itself, assembling the army to implement its mission. The signs of a computer virus are barely noticeable during the replication stage. However, if the virus starts spreading to other computers by sending infected files to everyone in your **contacts list**, your computer may run slower than usual.

3. **Trigger** event

A trigger event gives the virus a signal to act and bring the hacker's malicious intent to life. Depending on the virus type, anything can be a trigger, including a certain number of virus copies or a particular date.

4. Execution

This is the phase when the virus implements its mission. This can be anything from changing your **browser homepage** to stealing your passwords. During this stage, you may start noticing the damaging effects of a computer virus.

■ What does a Computer Virus do?

At this point, you might be wondering what kind of damage can computer viruses do? It all depends on the hacker's intentions, which may range from the desire to show off skills to terrorism. Viruses can delete programs, manipulate keyboards, **allow access to** sensitive information, or **flood** a network **with** traffic, making it impossible to do anything online. Can a virus destroy your computer? Yes.

Nevertheless, in the majority of cases, a computer virus is usually no more than a **nuisance** you can easily fix. But in some cases, viruses can lead to much more serious consequences. For example, the "I LOVE YOU" virus sent itself to 50 million users worldwide and added a password-stealing program to Internet Explorer, causing damage of up to $15 billion. And the "Sobig. F" virus stopped computer traffic in Washington DC, causing $37 billion in damages.

But are all computer viruses harmful? You may be pleasantly surprised to hear there is some sunshine among these dark clouds of evil intentions. The **case in point** is the Cruncher virus that can free up **hard drive space** by compressing every file it infects. There's also the Linux. Wifatch virus that functions as an antivirus. But unfortunately, "good" viruses are only a small subset of this software type.

■ Effects of Computer Viruses

Computer viruses are wholly dependent on their hosts. Just like their biological siblings, a virus doesn't alert you upon infection. Rather, it tries to remain unnoticed for as long as possible. So, what does a computer virus look like in terms of symptoms? Below are the most common signs.

1. Slower **operating speed**

Programs running in the background slow down a computer's speed. Since viruses are designed to run in the background and perform multiple actions, a performance **slump** is inevitable. If your Mac **takes ages to** start or open apps, then you might have caught a virus.

2. Issues with programs and files

Among the numerous hazards of computer viruses are data deletion and modification. Are there missing files? Are you unable to open certain apps? Do unknown programs start when you turn on your computer? Or, have new files, folders, or apps appeared on your hard drive **out of nowhere**? If so, **chances are that** malware has taken over your Mac.

Not unlike its biological **brethren**, a computer virus exists at the expense of a host, attaching itself to files and apps. Therefore, there's little surprise that a computer virus causes their noticeable modification.

The most common targets of viruses are system files. Without them, a computer system might not function correctly or even at all. Thus, by infecting these files, a **hacker** behind a virus can target the entire system **in a bid to** control your computer remotely.

3. Weird behavior

Just like the causes of a computer virus, its effects are only limited by the hacker's creativity. For example, after infecting the user's computer, the Elk Cloner virus displays a poem threatening to "stick to you like glue". And the Ika-Tako virus replaces all files, programs, and documents with pictures of **cuddly** squids.

If your **mouse pointer** starts jumping around the screen, don't blame it on a mischievous poltergeist. Most likely, it's either a drop of water on your **touchpad** or a computer virus.

4. Other effects

Certainly, poor performance, issues with programs and files, and weird behavior are not the only symptoms of infection by computer viruses. If people from your contacts list start receiving strange messages with attachments or **links** from you, it's probably the sign of a virus at work. If this does happen, then immediately change your passwords and ask everyone to delete those messages.

Another warning sign to look for is high network traffic, which often **spikes** during the self-replication stage of virus infection. Other signs of a virus include random browser redirects, a sudden lack of hard drive space, **system freezes**, or an unexplained **battery drain**.

(1092 words)

【Words, Phrases and Expressions】

parasite /ˈpærəsaɪt/　*n.* 寄生植物(或动物)，寄生虫

dormancy /ˈdɔːmənsi/　*n.* 休眠

trigger /ˈtrɪɡə/　*n.* (尤指引发不良反应或发展的)起因,诱因;引爆器,触发器

execution /ˌeksɪˈkjuːʃn/　*n.* 执行,实施

nuisance /ˈnjuːsəns/　*n.* 麻烦事,讨厌的人(或事物、情况)

slump /slʌmp/　*n.* 突然下跌;低迷期

brethren /ˈbreðrən/　*n.* 兄弟们;同胞

hacker /ˈhækə/　*n.* 黑客

cuddly /ˈkʌdli/　*adj.* 逗人喜爱的,可爱的

touchpad /ˈtʌtʃpæd/　*n.* 触摸屏设备,触摸板

link /lɪŋk/　*n.* (超文本)链接

spike /spaɪk/　*v.* 迅速提升,急剧增加

battery /ˈbætəri/　*n.* 电池,蓄电池

drain /dreɪn/　*n.* 流失,消耗

put sth. into perspective　客观地看待某物

in the same vein　同样地

contact list　联系人列表

browser homepage　浏览器主页

allow access to　允许访问

flood sth. with sth.　被……挤满,充满

case in point　恰当的例子

hard drive space　硬盘空间

operating speed　运行速度

take ages to do sth.　花很长时间去做……

out of nowhere　突然冒出来;莫名其妙出现

chances are that...　……是可能的

in a bid to　为了……

mouse pointer　鼠标指针

system freeze　系统死机

 Notes

1. host

A host (or network host) is a device that links with other hosts on a network. It can either be a client or a server that sends and receives applications, services, or data. Hosts have their unique IP address on a TCP/IP network, consisting of the device's local number and the network number it belongs to.

2. the Mac

It is short for the Macintosh, a personal computer made by Apple that has a graphical user interface (GUI) and a mouse. It was the first widely-sold computer with these features, which made it more user-friendly than previous computers. The Macintosh was launched in 1984 by Steve Jobs, the founder and CEO of Apple.

3. E-mail Hijacking

E-mail Hijacking, or e-mail hacking, is a widespread menace nowadays. It works by using the following three techniques which are e-mail spoofing, social engineering tools, or inserting viruses in a user computer.

4. "I LOVE YOU" virus

This virus name *I love you* referred to as a love bug or a love letter. It infected over 10 million computers and it started spreading as an e-mail message with the subject I love you. In that message, there is an attachment sent in the e-mail LOVE-LETTER-FOR-YOU. txt. vbs. Many users consider the. vbs extension as a plain text file. It catches many users' attention and makes them open the attachment. When the user opens the attachment, the Visual Basic script gets activated and damages the local machine. This virus is also capable of overwriting any files like images, audio, and it then sends a copy of it to all the addresses in the Windows Address book. It made the virus spread much faster than any spreading of e-mail worms.

5. Sobig. F virus

Sobig. F is the latest variant of yet another mass-mailing worm which exploits a vulnerability in the Microsoft Outlook e-mail client on Windows 95, 98, ME, NT, 2000, and XP platforms to replicate itself by mailing out messages with forged return addresses.

6. Cruncher virus

Cruncher is an unordinary virus; it stays resident in memory and infects files normally, when they are executed. In addition to this, it finally applies a self-extracting packing layer on the infected files, packing both the original code and the appended virus code.

7. Linux. Wifatch

Linux. Wifatch, which has been around since at least November 2014, uses Telnet and other protocols to hack into devices on which owners either set a weak password or left the default password unchanged. Once it infects a device, Wifatch scans it for known malware and disables Telnet to keep others out.

8. Elk Cloner virus

Elk Cloner is the first personal computer virus or self-replicating program known to have spread in the wild on a large scale. In 1982, 15-year old Richard Skrenta wrote the virus for the Apple II operating system. Stored on floppy disks, the virus copied itself to any uninfected floppy disks when a user booted a computer from an infected floppy disk.

9. Ika-Tako virus

The Ika-Tako virus (which is Japanese for Squid-Octopus) is a virus that runs on Microsoft Windows. It has started infecting in May of 2010, via Japanese file sharing Website Winny. It has infected between 20,000 to 50,000 computers. The virus disguises itself in music files, which users then download.

10. browser redirect

A browser redirect is when you are browsing online, a web page you are on causes your browser to redirect you to another page. The action is usually forced, unwanted, and often connected to malware, viruses, adware, browser hijackers, and ad-supported browser extensions.

 Exercise

A. Answer the following questions according to the text.

1. What is a computer virus and what distinguishes it from other malicious software?

2. When and why was a computer virus originally created?

3. What is the infection cycle of a computer virus?

4. How can viruses affect a computer according to the passage?

5. What symptoms might you be experiencing if your system has fallen victim to a computer virus?

B. Read the following statements and decide whether they are TRUE(T) or FALSE(F).

(　　)1. In the dormant phase, the virus won't self-replicate. Nor will it delete, capture or modify data on the infected computer.

(　　)2. The Sobig. F virus is one of the good viruses to help compress the files that it infects.

(　　)3. Viruses can delete and modify your data, infect your system files, or even destroy your computer.

(　　)4. Ika-Tako virus substitutes squids for all files, programs and documents due to the intention of the users.

(　　)5. An unexplained battery drain is another sign to indicate that your computer catches a virus.

Part 3　Extending Skill: Describing Graphic Information

It is considered an important academic skill to be able to intelligently and clearly interpret and describe data presented in graphs, charts and tables. In order to express information given in graphic forms, follow the three steps below.

■**Step 1: Identify Four Graphic Forms and Look for Trends**

① **Graph** (Figure 4 – 2)

A graph is a diagram containing lines or curves, which shows the trends of two or more sets of numbers or measurements. Here's an example on the right.

A graph plots the changes in data over time. You may focus on the following information:

· What is the highest level/point?

· What is the lowest level/point?

· Is there a point till when the trend was increasing or decreasing?

· When did the trend change?

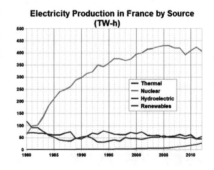

Figure 4 – 2

② **Pie charts (Figure** 4 – 3）

A pie chart is a circular chart divided into sectors or pie slices. It presents information in segments of a circle or pie, which together add up to 100％. Here's an example on the right.

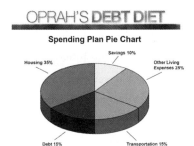

Figure 4 – 3

A pie chart is usually used to show percentages. You may focus on the following information:

- Which is the largest segment?
- Which is the smallest segment?
- How do the segments compare with each other?

③ **Bar charts (Figure** 4 – 4）

A bar chart is a diagram consisting of rectangular bars or columns arranged horizontally or vertically from the x or y axis. It makes information easier to be understood by showing the difference between two or more sets of numbers or measurements. Here's an example on the right.

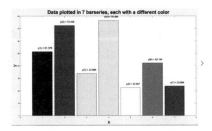

Figure 4 – 4

The length of a bar shows the values it represents. The values are listed on one axis and each bar shows what is being measured on the other axis. Bar charts are useful for comparing data. In studying a bar chart, you may want to look for the following information:

- Which is the tallest bar?
- Which is the shortest bar?
- Have the bars changed over time? How?
- How do the bars compare with each other?

④ **Tables**

A table is a set of facts and figures arranged in columns and rows. It is a very useful way of organizing numerical information. Figure 4 – 5 is an example.

Main Mode	Area of Work place					
	Central London	Rest of Inner London	Outer London	All London	Rest of Great Britain	Great Britain
Car and van	48	32	25	29	20	20
Motorbike, moped, scooter	36	29	27	31	19	21
Bicycle	33	24	20	25	15	17
Bus and coach	47	39	36	40	33	34
National Rail	69	66	43	66	47	58
Underground, tram, light rail	49	45	37	47	42	46
Walk	21	16	13	15	12	13
All modes	55	39	27	39	20	23

Figure 4 – 5

A table presents information in different categories, making it easy to compare. You may focus on the following information:

- What is the highest figure?
- What is the lowest figure?
- What is second highest?

■Step 2: Vocabulary of Trends

In order to provide precise descriptions of the graphic information, a range of vocabulary is needed:

- Verbs to describe downward movement: decline, decrease, drop, fall, crash, slide, dive, collapse, plunge, plummet;
- Verbs to ddescribe upward movement: grow, soar, increase, rocket, rise, climb, surge, jump;
- Verbs to describe stability: remain at, hold steady, even off, is unchanged, stay the same, stabilize, level off;
- Adjectives to specify the pace or degree of changes: slow, massive, gradual, remarkable, slight, rapid, moderate, sharp, marginal, dramatic, steady, significant;
- Adverbs to specify the pace or degree of changes: slowly, massively, gradually, remarkably, slightly, rapidly, moderately, sharply, marginally, dramatically, steadily, significantly.

■Step 3: Write According to the Instruction

When describing the graphic information, you should write according to the following instruction:

- Firstly, introduce the subject or the graphic.
- Secondly, show the relationship between the data or to comment on the main trend.
- Thirdly, give an in-depth description of the graphic information. Include details from the most important to the least important, depending on time.

Remember that in this section you should only focus on the data. Don not give your interpretation of it. You are not being asked why you think the data in the chart is the way it is, so do not write about that.

Example 1

Directions: In this section, you are required to describe the following diagram.

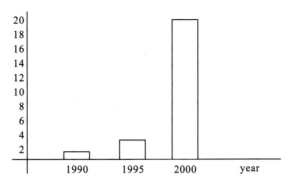

Average Number of Hours a Student Spends on the Computer per Week

(2002. 6 CET-4)

From the chart we can see clearly that the average number of hours a student spends on the computer per week increased dramatically. It rose from less than two hours in 1990 to nearly four hours in 1995. In 2000, the number soared to 20 hours.

Example 2

Directions: In this section, you are required to describe the following diagram.

Reading Habits of American Students

(2003. 9 CET-6)

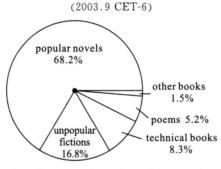

It can be seen from the diagram that American students prefer to read popular novels rather than other kinds of books. According to the statistics above, the popular novels occupy 68.2 percent of all. While unpopular fictions make up 16.8%, technical books account for 8.3%, poems only 5.2% and other books 1.5%.

Example 3

Directions: In this section, you are required to describe the following table.

Changes in People's Diet

Food Year	1986	1987	1988	1989	1990
Grain	49%	47%	46.5%	45%	45%
Milk	10%	11%	11%	12%	13%

续表

Food Year	1986	1987	1988	1989	1990
Meat	17%	20%	22.5%	23%	21%
Fruit and Vegetables	24%	22%	20%	20%	21%
Total	100%	100%	100%	100%	100%

(1991.6. CET-4)

This table shows that there were great changes in people's diet from the year 1986 to 1990. According to the data above, the consumption of grain significantly declined from 49% to 45%. It was also the case with the consumption of fruit and vegetables. However, during the same period the proportion of milk and meat in people's diet was on the increase.

【Exercise】

Here is a diagram. You are required to describe the diagram by using the methods of this unit.

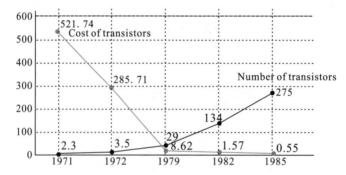

Part 4 Optional Exercises

A. Translate the following paragraph into Chinese.

As you may know, a biological virus is a tiny parasite relying on living cells for survival. Viruses live at the expense of the host, which can be an animal, plant, or bacteria. As soon as it has infected a host cell, a virus starts reproducing, invading other cells, and spreading disease. In the same vein, a computer virus shows up on your Mac uninvited and starts reproducing itself. If left unchecked, it can cause a lot of damage. Unable to function without a host, this type of malware travels from computer to computer, program to program, file to file.

B. Translate the following paragraph into English.

计算机病毒在二十世纪七十年代初首次被创造出来,造成了不同程度的影响。有些会使文件被删除,有些则在电脑一开机就显示特定的信息。今天的计算机病毒比以前蔓延得更广也更危险。到目前为止,最具破坏性的病毒也许就是"我爱你"病毒了。"我爱你"病毒于 2000 年 5 月首次出现,就金钱方面的损失而言,它也许是毁坏性最强的病毒。当包含该病毒的电子邮件附件被打开时,病毒就会被释放出来。通过更改计算机中的文件名,"我爱你"病毒使计算机难以读取文件。它也会搜寻如密码等类型的重要的个人数据,将其传送到网站上让其他人看到。

C. Write a composition on the topic "Information Security". You should write at least 160 words but no more than 200 words following the outline given below in Chinese.

　　a.信息安全问题日益重要。

　　b.信息安全事故可能导致的危害。

　　c.如何做到信息安全?

Unit 5

E-commerce

Part 1   Vocabulary

A. Choose the full names in Column B that best match the terms in Column A.

Column A	Column B
1. EDI	a. Hypertext Marked Language
2. e-commerce	b. merchant
3. HTTPS	c. shopping basket
4. FTP	d. electronic data interchange
5. revenue	e. a malware
6. SSL	f. electronic commerce
7. shopping car	g. File Transfer Protocol
8. HTML	h. income
9. dealer	i. Hypertext Transport Protocol
10. phishing	j. Source Socket Layer

B. Use the words from Exercise A to label the following pictures.

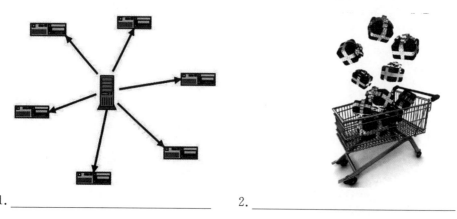

1.＿＿＿＿＿＿＿＿＿＿＿＿＿＿＿   2.＿＿＿＿＿＿＿＿＿＿＿＿＿＿＿

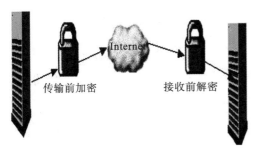

3. _____

4. _____

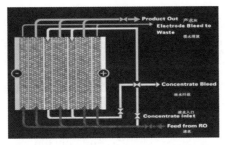

5. _____

6. _____

C. Complete the following sentences with the words given below. Change the form if necessary.

transaction	categorize	purpose	dealer	primarily
physical	execute	fraudulent	repudiation	countermeasure
antivirus	manipulate	feed	oversight	calculate

1. The _____ of the occasion was to raise money for medical supplies.

2. The advertisers just stopped making payments to the _____ websites.

3. The framework will automatically locate and _____ files for you, without requiring any manual intervention.

4. Online purchases can be made by setting up a separate _____ code.

5. William was angered and embarrassed by his _____.

6. Sometimes there are not many choices in a _____ shop, but we can always find an infinite variety of goods online.

7. As hackers turned up, layers of security, from _____ programs to firewalls, were added to try to keep them at bay.

8. As Mercedes Benz 4S _____ didn't handle the oil leak scandal, we have seen its terrible consequences.

9. Technorati uses tags to _____ blog article.

10. The body is made up _____ of bone, muscle and fat.

Part 2   Reading

 Text A

Application and Safety of E-commerce Business

Electronic commerce, commonly known as e-commerce or e-comm., refers to the buying and selling of products or services over electronic systems such as the Internet and other computer networks.

Electronic commerce **draws on** such technologies as electronic funds transfer, supply chain management, Internet marketing, online transaction processing, electronic data interchange (EDI), **inventory** management systems, and automated data collection systems. Modern electronic commerce typically uses the World Wide Web at least at one point in the transaction's **life-cycle**, although it may **encompass** a wider range of technologies such as e-mail, mobile devices and telephones as well.

Different e-commerce activities have different purposes. Is the business targeting consumers like you or does it cater to another business? Maybe the e-commerce isn't even a company, but someone trying to sell something to someone else on the Web. Some people find it useful to categorize e-commerce by the types of **entries** participating in the **transaction** or business processes. The five general e-commerce **categories** are business to consumer, business to business, business process, consumer to consumer, and business to government. The three categories are most commonly used are:

Business to Consumer E-Commerce (B2C)

Business to Consumer electronic commerce (B2C e-commerce) occurs when a business sells products and service through e-commerce to customers who are primarily individuals. Hundreds of thousands of B2C e-commerce business exit on the web. The most famous B2C WEB site is Taobao (www.taobao.com), Jingdong, (www.360buy. com), Amazon (www.amazon.cn), etc. B2C e-commerce businesses have been the most visible. However, only about 3 percent of all e-commerce revenues are from B2C. The other 97 percent are mostly B2B e-commerce revenues.

Business to business E-commerce (B2B)

Business to business electronic commerce (B2B e-commerce) occurs when a business sell product and service through e-commerce customers who are primarily other businesses. For example, Gates Rubber Company makes rubber and synthetic-based product such as belts and hoses for car. But Gates doesn't sell directly to you. Instead, parts dealer and automakers buy its products.

Consumer to Consumer E-Commerce (C2C)

Consumer to Consumer E-Commerce (C2C e-commerce) occurs when a person sells products and service to another person through e-commerce. The most well known Web site that supports C2C e-commerce is 58. com (www. 58. com), baixing (www. baixing. com), eBay(www. ebay. com), etc.

Business to Government E-commerce (B2G)

Business to Government E-commerce includes business transactions with government agencies, such as paying taxes and filling required reports. An increasing number of states have web sites that help companies do business with state government agencies.

Online shopping is a type of E-commerce used for B2B and B2C transactions. Consumers find a product of interest by visiting the website of the **retailer** directly, or do a search across many different **vendors** using a shopping search engine. Once a particular product has been found on the website of the seller, most online retailers use shopping cart software to allow the consumer to accumulate multiple items and to adjust quantities, by **analogy** with filling a physical **shopping cart** or basket in a conventional store.

Today shopping cart are a standard of e-commerce. A shopping cart (Figure 5 – 1) also sometimes called shopping bag or shopping basket, keep track of the items the customer selected and allows customer to view the contents of their carts, add new items or remove items, to order an item, the customer simply click the item. All of the details about the item, including its price, product number, and other identifying information, are stored automatically in the cart. If the customer changes his mind about the item, he can view the cart's contents and remove the unwanted items. When the customer is ready to conclude the shopping session, the click of a button **executes** the purchase transaction.

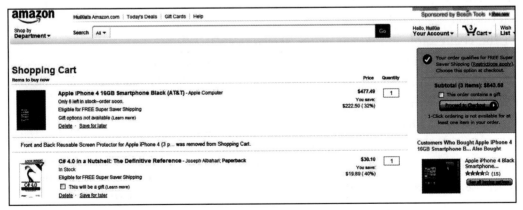

Figure 5 – 1   A Typical Shopping Basket Page at an E-Commerce Site

Clicking the **checkout button** usually displays a screen that asks for billing and shipping information and that confirms the order. The shopping cart software keeps a running total of each type of item. The shopping cart calculates a total as well as sales tax and **shipping costs**.

Given the lack of ability to inspect **merchandise** before purchase, consumers are at higher risk of **fraud** on the part of the merchant than in a physical store. Merchants also risk fraudulent purchases using stolen credit cards or **fraudulent repudiation** of the online purchase.

Source Socket Layer(SSL) encryption has greatly solved the problem of credit card numbers being **intercepted** in transit between the consumer and the **merchant.** Identity theft is still a concern for consumers when hackers break into a merchant's web site and steal names, addresses and credit card numbers. Computer security has this become a major concern for merchants and e-commerce service providers, who deploy countermeasures such as **firewalls** and **antivirus** software to protect their networks.

Phishing is another danger, where consumers are fooled into thinking they are dealing with a reputable retailer, when they have actually been **manipulate** into **feeding** private information to a system operated by a malicious party. To protect your private information not be stolen and better keep safety process of your e-commerce, you'd better follow the suggestions as below:

Sticking with known stores, or attempting to find independent consumer reviews of their experience, also ensuring that there is **comprehensive** contact information on the website before using the service, and noting if the retailer has enrolled in industry **oversight** programs such as trust mark or trust seal.

Ensuring that the retailer has an acceptable privacy policy posted. For example, note if the retailer does not explicitly state that it will not share private information with others without **consent**.

Ensuring that the vendor's address is protected with SSL when credit card information is entered. If it is, the address on the credit card information entry screen will start with "HTTPS".

Using strong passwords without personal information. Another option is a "**passphrase**", which might be something along the lines:"I shop 4 good a buy!!"It is difficult to hack, and providers a variety of **uppercase, lowercase**, and special characters and could be site specific and easy to remember (Figure 5 – 2).

Figure 5 – 2　Interface of Login in the Typical E-Commerce Web Site

(1001 words)

【Words，Phrases and Expressions】

inventory /ˈɪnvəntəri/   *n.* 存货清单；物品清单；财产清单；

life-cycle /laɪf ˈsaɪkl/   *n.* 生命周期；生存期；

encompass /ɪnˈkʌmpəs/   *vt.* 围绕；包围；包括；完成

entry /ˈentri/   *n.* 入口；条目

transaction /trænˈzækʃn/   *n.* 交易，买卖，业务；(学术团体会议的)议事录，公报

category /ˈkætəgəri/   *n.* 分类

revenue /ˈrevənjuː/   *n.* 税收；收入

retailer /ˈriːteɪlə/   *n.* 零售商

vendor /ˈvendə/   *n.* 自动售货机；小贩；卖方；供货商

analogy /əˈnælədʒi/   *n.* 类比，比拟

execute /ˈeksɪkjuːt/   *vt.* 执行；实行；处决；完成

merchandise /ˈmɜːtʃəndaɪs/   *n.* 商吕，货品   *v.* (用广告等方式)推销(商品或服务)

fraud /frɔːd/   *n.* 欺骗；欺诈；诈骗罪；骗子；赝品；仿造品

repudiation /rɪˌpjuːdɪˈeɪʃən/   *n.* 否认；拒付；拒绝承认

content /kənˈtent/   *n.* 内容；意义；(网站、光盘等上的)电子信息，存储信息；含量；成分
                       adj. 满意的；满足的   *vt.* 使满意；使满足

intercept /ˌɪntəˈsept/   *vt.* 截击；拦截；截住；截获；拦截；阻截；拦截；截取；截断

merchant /ˈmɜːtʃənt/   *n.* 商人，(尤指外贸)批发商

countermeasure /ˈkaʊntəmeʒə(r)/   *n.* 对策

firewall /faɪrwɔːl/   *n.* 防火墙

antivirus /ˈæntɪvaɪrəs/   *n.* 反病毒程序

manipulate /məˈnɪpjʊleɪt/   *vt.* 摆布；操纵；处理；篡改

feed /fiːd/   *vt.* 喂养；饲养；靠……为生；向……提供   *vi.* 吃饲料；进餐   *n.* 一餐；饲料

comprehensive /ˌkɒmprɪˈhensɪv/   *adj.* 广泛的；无所不包的；全面的；彻底的；完全的

oversight /ˈəʊvəsaɪt/   *n.* 失察；失误；疏忽；监督

consent /kənˈsent/   *n.* 许可，允许；同意，赞同   *vt.* 赞成，准许

passphrase /pɑːsˈfreɪz/   *n.* 口令

uppercase /ˈʌpəˌkeɪs/   *n.* 大写   *adj.* 大写的

lowercase /ˈləʊəˌkeɪs/   *n.* 小写   *adj.* 小写的

draw on   利用

shopping cart   购物车

shipping costs   运费

checkout button   购买按钮

fraudulent repudiation   欺诈

Source Socket Layer(SSL)   源层

Notes

1. SSL

SSL is a process undergone by data under the SSL protocol in order to protect that data during transfer and transmission by creating a channel, uniquely encrypted, so that the client and the server have a private communication link channel over the public Internet. This is how encryption protects data during transmission.

2. HTTPS

HTTPS is the lock icon in the address bar, an encrypted website connection—it's known as many things. While it was once reserved primarily for passwords and other sensitive data, the entire web is gradually leaving HTTP behind and switching to HTTPS. The "S" in HTTPS stands for "Secure". It's the secure version of the standard "hypertext transfer protocol" your web browser uses when communicating with websites.

 Exercise

A. Answer the following questions according to the text.

1. What technologies are likely to be used in e-commerce?

2. According to the types of entries participating in the transaction, what are the types of e-commerce?

3. Can you give an example of C2C type?

4. What technology can solve the credit card security problem in e-transations to some extent?

5. What should we do to protect our personal information when shopping online?

B. Choose the best answer to each of the following questions according to the text.

1. Which of the following statements about e-commerce is NOT true?

   A) E-commerce is short for electronic commerce.

   B) E-commerce refers to the buying and selling of products or services over the Internet and other computer networks.

   C) Online shopping is a type of E-commerce used for G2C transactions.

   D) Different e-commerce activities have different purposes.

2. The following activity is _____.

   George sold the bike he had just bought on ebay.

   A) B2C   B) B2B   C) C2C   D) G2C

3. Which of the following is a typical B2C website?

    A) Taobao   B) 58.com   C) eBay   D) baixing

4. Which is NOT part of the shopping cart function?

    A) It can keep track of the item.

    B) It can add new items or remove items.

    C) It can allow customer to check the contents of their cart.

    D) It can calculate a total as well as sales tax and shipping costs automatically.

5. It can be inferred from the passage that _____ .

    A) e-commerce must be a company or a website.

    B) e-commerce may be someone trying to sell something to someone else on the Internet

    C) with SSL, consumers needn't worry about the online security

    D) phishing is software which is used to protect online-shopping safety

 Text B

E-commerce Development

Electronic Commerce is exactly **analogous** to a marketplace on the Internet. Electronic of products or services over electronic systems such as the Internet and other computer networks. The information technology industry might see it as an electronic business **application** aimed at commercial transactions; in this **context**, it can involve electronic funds **transfer**, supply **chain** management, e-marketing, online marketing, online transaction processing, electronic data interchange (EDI), automated **inventory management** systems, and automated data collection systems. Electronic commerce typically uses electronic communications technology of the World Wide Web, at some point in the transaction's lifecycle, although of course electronic commerce frequently depends on computer technologies **other than** the World Wide Web, such as databases, and e-mail, and on other non-computer technologies, such as transportation for physical goods sold via e-commerce.

Since the ninety's, the electronic commerce in the context of the global rise and rapid development, quickly change the **original** economic pattern, and traditional economic operation and growth mode. The development of e-commerce has become a key economic development **evaluation** of comprehensive strength important index. Therefore, understanding our country electronic commerce development present situation, the **objective** understanding to the problems of China's e-commerce development, the countermeasures to achieve the fast, healthy and stable development, has become a new era of the economic development of the construction of **harmonious** society an urgent and important tasks.

Companies have conducted online **commercial** transaction on a limited **scale** since the early 1990s. Consumer's buying, however, has had a different history. Nowadays, consumers can count on one hand the years for which significant numbers of consumer have been buying products online. Almost daily, Internet developers design new technologies that business operations and automate the purchasing process. The goal of all this rapid development is to attract more consumers by making Internet buying and selling **procedures** as fast, convenient, and easy as possible. In the following, the key **driver** is of E-commerce development:

Internet resources

The Internet consists of a vast **array** of electronic resources that people use to access information, to communicate with each other, and transmit data. Sometimes people use the words Internet and Word Wide Web **interchangeably** to refer to all these electronic resources. Internet describes entire system of network computers; Word Wide Web describes the method used to access information contain on computers related to the Internet.

By the mid-1990s, companies and individuals began to recognize the **potential** of the Web for reaching both exiting and new customers. Business in particular **identified** the potential cost saving they could realize by using the Internet handle business transactions. To start with, the Internet is **profoundly** changing consumer behavior. One in five customers walking into a Sears department store in America to buy an electrical appliance will have researched their purchase online—and most will know down to a dime what they intend to pay. More surprisingly, three out of four Americans start shopping for new cars online, even though most end up buying them from traditional dealers. The difference is that these customers come to the showroom armed with information about the car and the best available deals. Sometimes they even have computer print-outs identifying the particular vehicle from the dealer's stock that they want to buy.

Intense competition

Intense competition , proliferation of products service and high consumer expectations in nearly business have added unusual pressure to keep close watch on operation costs and maximize profit margins. In order to survive, companies are constantly looking for more effective ways. E-commerce addresses these concerns quickly, efficiently, and at low cost.

Globalization

To maintain growth of profit, many companies are moving to the international market. However, one of the major obstacles is the **geographical** barrier. E-commerce provides an effective "vehicle" for companies to move to the international market because there is almost no geographical barrier in cyberspace. In other words, it is easier for a foreign company to compete with a local company under the cyber environment.

E-mail and FTP

Other terms associated with the Internet are e-mail and File Transfer Protocol. People use e-mail to send messages from one computer to another computer. People can use FTP transfer documents and files from one computer to another via the Internet. The **World Wide Web**, e-mail, and FTP are all tools that people use to access information from the Internet.

Automation

As the cost of labor increase, there is a strong need for companies to look for **alternative** ways to do routine work. This is particularly true in handing the myriad paper transaction once an order is taken. With electronic messages one can reduce this **considerably**; e-commerce thus provides an attractive solution.

Nowadays, e-commerce in China's development has a certain scale, especially since 2006 years of the electronic commerce in our country is developing rapidly. E-commerce led the whole economic development in China; the economic development of our country has entered a new stage. But, as our country's geography, economic policy, the influence of factors such as the development of regional economy. E-commerce in promoting the development of the economy as a whole to regional economic development also brings opportunities and challenges. In the underdeveloped regions, and make good use of it, is the best opportunity for himself, and to create a competitive advantage, can take up to rapidly developed regions. Ignore it, may continue to lag behind or lose advantage may result from the developed areas. 　　(895 words)

【Words, Phrases and Expressions】

analogous /əˈnæləgəs/　*adj.* 相似的;类似的;可类比的;可比拟的;同功的;功能相同但起源不同的;类推出的

application /ˌæplɪˈkeɪʃn/　*n.* 应用软件;应用;申请

context /ˈkɒntekst/　*n.* 上下文;语境;情况,环境,背景

transfer /trænsˈfəː/　*vt.* 转移;调动;转存

　vi. 转移;迁移;移动;移交;转乘;换乘

　n. 转账;移交;转移者;(财产、权利的)转让,让与;(证券资产的)转让;过户

chain /tʃeɪn/　*n.* 链;链条;项链;束缚;羁绊;限制;一连串,一系列;连锁店

original /əˈrɪdʒənəl/　*adj.* 起初的;最初的;原先的;原件的;原作的;原创的;独创的;有独创性的

　n. 原件;原作;(艺术或文学作品的)原型

evaluation /ɪˌvæljuˈeɪʃən/　*n.* 评估;评价;量化;计算;求值

objective /əbˈdʒektɪv/　*n.* 目的;目标;宗旨;出击目标;(光学仪器的)物镜;(照相机或投影仪成像的)镜头

harmonious /hɑːˈməʊnɪəs/ *adj.* 和谐的；协调的；调和的；友好的；和谐的

commercial /kəˈmɜːʃəl/ *adj.* 商业的；贸易的；商务的；营利性的；商业化的；商品化的；供出售的；以营利为目的的

scale /skeɪl/ *n.* (相对的)大小，规模，范围；尺度 *vt.* 爬越；攀登

procedure /prəʊˈsiːdʒə/ *n.* 步骤；过程；手续；程序；方式

driver /ˈdraɪvə/ *n.* 驾驶员；司机；私人司机；牧人；驱动程序；原动力；推动力

array /əˈreɪ/ *n.* 一系列，大量；数组；陈列

interchangeably /ˌɪntəˈtʃeɪndʒəbli/ *adv.* 可交换地；可互换地；可交替地

potential /pəˈtenʃəl/ *adj.* 潜在的；潜在的；可能的；表示可能性的
n. 潜力；潜能；潜质；可能性

identify /aɪˈdentɪˌfaɪ/ *vt.* 把(自己)与……密切联系；认为(自己)与……有关联
vi. 认同；产生共鸣

profoundly /prəˈfaʊndli/ *adv.* 深刻地；深奥地；深深地

geographical /ˌdʒiːəˈɡræfɪkl/ *adj.* 地理(学)的；地理特征的；地理的

alternative /ɔːlˈtɜːnətɪv/ *adj.* 可供选择的；可替代的；另类的；非主流的；非传统的；(尤指)两者择一

considerably /kənˈsɪdərəbli/ *adv.* 相当地

inventory management   库存管理
other than   不同于；除……以外

Notes

1. EDI

EDI is short for Electronic Data Interchange, the transfer of data between different companies using networks, such as VAN's or the Internet. Electronic Data Interchange (EDI) is the process of using computers to exchange business documents between companies. EDI is used by a huge number of businesses. Over 100,000 businesses have replaced the more traditional methods with EDI.

2. The File Transfer Protocol (FTP)

FTP is a communication protocol providing for the transfer of files between remote devices and a server across a local (LAN) or wide-area network (WAN) like the Internet. FTP servers and software were crucial to the development of early networking in the last quarter of the 20th century and served as a precursor to more secure alternatives like FTPS and SFTP. Today, FTP has become less popular with cloud storage and sharing.

Exercise

A. Answer the following questions according to the text.

1. In the field of e-commerce. What role may information technology industry serve as?
2. What information technology may be involved in e-commerce.
3. What are the factors that promote the development of e-commerce?
4. What is the major obstacle when companies turn to international markets?
5. What tools people can use to acquire information from the Internet?

B. Read the article and decide whether the following statements are TRUE(T) or FALSE(F).

(　　) 1. Since the 1990s, e-commerce has emerged and developed rapidly all over the world.

(　　) 2. People mainly use the Internet to have access to electronic resources.

(　　) 3. E-commerce effectively solves the pressures faced by enterprises, such as fierce competition, product proliferation and high expectation from consumers.

(　　) 4. In the context of globalization, the Internet makes it impossible for foreign enterprises to compete with local ones.

(　　) 5. The automation feature of e-commerce saves the labor cost of enterprises.

Part 3　Extending Skill: Search Engines

There are several popular search engines like 360, Baidu. Using search engine is not difficult in the least. However, there is a way of using a search engine properly. You want to increase the results you get from search engines to make sure they all are related to the subject you are researching. Here is how you can properly use a search engine.

■ Necessary Items

　　Computer and Internet access

■ Instructions

　　Step 1

　　When you need to research a phrase that has more than one word in it there is a better way of getting the right results. First, you should place the phrase in quotation marks. This signifies that you need an exact match to wording. Also, you can add the word "and to the phrase to help. The search engine will search the Internet for your keyword and populate your screen with results. The results that come up are websites that contain the subject matter requested.

Step 2

Using quotation marks around a word or phrase will narrow the search down. You can do this when you are looking up things like music lyrics, famous quotes and more. The results will be an exact match to the keyword you entered at the beginning of the search. All the websites listed will have the phrase on the site.

Step 3

Make sure the spelling and order of your words are correct. Spelling words differently can affect the results of your search, especially words that are transliterated from another language. Recheck all spellings to ensure you get accurate results. Some search engines will automatically correct the spelling of certain words and ask if you mean that phrase. You will see this populate along with your results at the very top of the options. The phrase is clickable. So, if the correction is accurate you don't need to conduct a separate search, all you need to do is click that phrase. The live link will repopulate more accurate results for you and you may just from the options provided. If you use these techniques you will be able to better use a search engine to get the results you need.

【Exercise】

A. You want to research the following. Choose up to three keywords or phrases for each search.

1. Sales on Tmall's Single Day in 2021.
2. Requirements of JD(京东) to set up online shops.
3. How does online shopping safeguard its rights?

B. Analyze Internet search results and answer the following questions.

1. Where can we find the website address in each result?
2. Which are PDF documents?
3. Which results contain abbreviations or acronyms?
4. Which country does result 7 come from?
5. Which results come from web service companies commercial ?

Bai**百度** | the future of e-commerce in China 　　　　　　　　× ◎ | **百度一下**

Q 网页　　□ 文库　　国 资讯　　贴 贴吧　　⑦ 知道　　□ 图片　　凸 地图　　回 采购　　▷ 视频　　**更多**

1　https://www.economist.com › weeklyedition　⋮

The future of e-commerce (with Chinese characteristics)

Jan 2, 2021 — **The future of e-commerce** (with **Chinese** characteristics) – Weekly edition of The Economist for Jan 2nd 2021. You've seen the news, ...

2　https://www.tradegecko.com › hubfs › eBooks 〔PDF〕　⋮

What China reveals about the future of eCommerce

China's **eCommerce** market was valued at $750 billion in 20161, making it the largest market in the world and worth more than the US and UK combined.
13 pages

3　https://www.researchgate.net › publication › 29738726_T...　⋮

(PDF) The future of e-commerce in China - ResearchGate

Jul 5, 2022 — **China** has the resources, the means, and the motivation to be a central player in the global **e-commerce** industry but is lagging far behind ...

4　https://www.shopify.ca › global-ecommerce-statistics　⋮

Global Ecommerce: Stats and Trends to Watch (2022) - Shopify

Feb 16, 2022 — After **China** and the US, the third-largest **ecommerce** market is the United Kingdom, taking up 4.8% of the retail **ecommerce** sales share. The UK is ...

5　http://jingdaily.com › downloads › the-future-of-cross-...　⋮

The Future of Cross-Border E-Commerce in China - Jing Daily

May 17, 2022 — **E-commerce** is a global industry with global opportunities, and **China**, with its sizable market and growing population of affluent consumers, ...

6　https://www.retail-insight-network.com › analysis › e-c...　⋮

E-commerce in China: A market overview

Oct 7, 2019 — "The growth of **China's e-commerce** has been spurred on by wider socio-economic factors. Early **e-commerce in China** tended to focus on low value, ...

7　https://www.irishtimes.com › business › technology › how...　⋮

How China is shaping the future of shopping - The Irish Times

May 13, 2021 — Fast forward to 2021 and live video has become a core feature of **China's** biggest **ecommerce** apps such as Taobao Live, JD.com and Pinduoduo.

8　https://www.globaldata.com › chinese-e-commerce-mar...　⋮

Chinese e-commerce market to reach US$3.3 trillion in 2025 ...

Sept 14, 2021 — **Chinese e-commerce** market, the world's largest, is estimated to register a strong growth of 17.2% in 2021, as consumers increasingly shift ...

9　https://ecommerceguide.com › guides › social-ecommer...　⋮

Social Ecommerce: The Future of Online Shopping is in China

Olivier Verot has been living in **China** for almost 8 years and founded a digital strategy agency specialized in the **Chinese** market -

10　https://www.most2414.com › 5-trends-future-of-ecom...　⋮

5 Trends for the Future of E-commerce in China - MOST 2414

More than 40% of online sales transactions are in **China**... and just ten years ago it was 1%. ValueChina reports what can be 5 trends of **e-commerce in China**.

Part 4 Optional Exercises

A. Translate the following paragraph into Chinese.

To maintain growth of profit, many companies are moving to the international market. However, one of the major obstacles is the geographical barrier. E-commerce provides an effective "vehicle" for companies to move to the international market because there is almost no geographical barrier in cyberspace. In other words, it is easier for a foreign company to compete with a local company under the cyber environment.

B. Translate the following paragraph into English.

如今,很多人都会借助互联网在家里舒舒服服地购物,网络购物已成为人们最喜爱的购物方式之一。对于消费者来说,网络购物不仅方便、选择范围广、价格具有竞争性,而且更容易获取商品信息。对于商家来说,网络提供了更多的客户和更大的市场空间。对于整个市场经济来说,这种新鲜的购物模式可在更大的范围内及更广的层面上以更高的效率实现资源配置(allocate resources)。

C. Write a composition on "Online Shopping", basing on the outline given below.

a. The popularity of online shopping

b. The advantages and disadvantages of online shopping

c. Your opinion about online shopping

Unit 6

Computer Assisted Education

Part 1   Vocabulary

A. Choose the full names in Column B that best match the terms in Column A.

Column A	Column B
1. CAL	a. computer-aided design
2. CALL	b. human computer interaction
3. CAD	c. compact disc
4. VLEs	d. Moving Picture Experts Group Audio Layer III
5. HCI	e. computer assisted learning
6. Edtech	f. virtual learning environments
7. DVD	g. education technology
8. CD	h. Computer Assisted Language Learning
9. MP3	i. Electronic Learning
10. E-learning	j. Digital Versatile Disc

B. Use the words from Exercise A to label the following pictures.

1. _____   2. _____

3. _____   4. _____

5. _____   6. _____

C. Complete the following sentences with the words given below. Change the form if necessary.

distraction	incorporate	encompass	heighten	summarize
implement	personalize	navigate	disposal	emulate
privacy	retain	target	associate	effective

1. I do very much appreciate the quietness and _____ here.

2. The tradition, dating back to 300 B.C., was later _____ into the Christian church.

3. The course will _____ physics, chemistry and biology.

4. It will also _____ fears over inflation, which remains stubbornly high.

5. If I can be of service, I am at your _____.

6. He has to learn to _____ by electronic instruments.

7. Sons are traditionally expected to _____ their fathers.

8. Recently I attended several meetings where we talked about ways to _____ students and keep younger faculty members from going elsewhere.

9. Foreigners always _____ China with the Great Wall.

10. Thanks to recent research, _____ treatments are available.

Part 2　Reading

 Text A

What is Computer Assisted Learning and How Does It Work?

In today's classroom setting, it seems like we're constantly battling with technology.

On one hand, technology has **revolutionized** the way teachers teach and students learn. On the other hand, it can become a **distraction** that takes away from the learning process. If you're a teacher like me, you know that it can be a huge pain to **get** students **off** their smartphones or computers and get them to **pay attention to** the lesson. All you can see are the tops of their heads while they **text** and **post on** social media—and you just know they aren't posting about the **irregular** verbs you're currently teaching them!

Luckily, there is a way to **incorporate** that very same technology that is distracting your students **into** your classroom environment and use it to help them learn better.

■ **What is Computer Assisted Learning?**

Computer Assisted Learning encompasses a lot of different technologies and ideas, but can be understood easily enough. The Intense School, which focuses on computer and information technology, **summarizes** it simply as "the use of electronic devices/computers to provide educational instruction and to learn."

It might shock you to learn that some form or another of Computer Assisted Learning has been taking place in classrooms since the 1960s. CAL doesn't just involve computers, it also includes the use of other electronics such as CD and MP3 players (or record players in the 1960s), DVD players, **tablets**, smartphones and televisions. These tools can be used to better illustrate a point the teacher or professor is trying to make, or to **heighten** engagement among students.

Just think about it: Wouldn't you learn more from actually watching a foreign film for your language class than you would from just talking about it?

Computer Assisted Learning also includes online courses and **supplemental** course materials used in colleges, homeschooling and distance learning. Basically, any type of technology that can be used to learn most likely falls beneath the umbrella of Computer Assisted Learning.

The great thing about this teaching method is that it can be **implemented** in every type of classroom, from a kindergarten art class to a medical school class in which students use computer models to learn to **operate on** the human body. It can also help students take classes at home either on their own or to supplement their other learning. This can lead to a much more **personalized** experience, as well as a more in-depth understanding of the knowledge being transmitted.

■ How Can Computer Assisted Learning Help Students with Languages?

While the use of CAL can be useful in any classroom, it's especially beneficial in language learning classrooms.

In fact, it's so effective that it gets its own **acronym** too! CALL, or Computer Assisted Language Learning, is quickly becoming one of the preferred teaching tools among foreign language instructors.

Using CALL, language teachers can help their students retain more vocabulary and grammar by having them watch videos, play computer games, or even **navigate** the Internet using only their target language. It also enables students to use that target language in a more active way, which helps them learn it more naturally than just **rote memorization**. The words and rules of the language become something more useful to them, so they're able to remember them better.

Here are just a few examples of how Computer Assisted Learning can be used to help students learn languages:

Visual Learning

Many students are visual learners, and benefit greatly from seeing an image or an example of the terms being discussed in class. Computers are a great help with this, because teachers have the entire Internet **at their disposal**. You can easily search the web for pictures of fruits, animals or even colors to help your students see what you mean and have an image to **associate with** the word you're describing.

You can also use videos from DVDs, YouTube or your own personal projects to help illustrate a point. Seeing something really happening or really being used in a video makes it much more real to the learner, so they remember it much longer.

Listening Practice

Listening practice is a vital part of learning any language. CAL helps with this by enabling you to play music or record conversations, so your students can listen to the language being used naturally and in real situations. They can then **emulate** the speakers or singers and find their own voice in their new tongue.

Tests

Computers are a great way to give students exams. You can either create your own test and have them sit at the classroom computers to take it, or you can find pre-written tests and other exam materials on the Internet and use those in your lessons. Taking tests on the computer can help students feel less rushed and can make them feel as if they have more privacy than they would if they were in a crowded classroom.

Games

Games are perhaps one of the best ways to use CAL in the classroom. Language students (especially young ones) love playing computer games or doing puzzles in their **target** language.

To them, it doesn't feel like learning—it feels like having fun. They won't even realize they are getting smarter as they try to get to the next level or solve a tough crossword, when in fact, they're learning and retaining more than they would have otherwise!

Internet Searches

Another fun way to use a student's target language in the classroom is by having them do an Internet search in said language. Activities like WebQuest begin with the teacher giving students a **query** to look up with a search engine. The students then have to find the answer using only their target language, which can be a real (but fun!) challenge!

Online Courses

Last but certainly not least, CAL can include online courses. These courses can be **taken on** one's own time at home, possibly as a part of a full college course load, or they can be taken as a supplement to a language course they are already taking in person. There are hundreds of free or paid language courses to be found online, and many of them can be extremely effective.

(1032 words)

【Words, Phrases and Expressions】

revolutionize /ˌrevəˈluːʃənaɪz/   *vt.* 发动革命；彻底改革；使革命化

         vi. 革命化；引起革命

distraction /dɪˈstrækʃn/   *n.* 使人分心的事；娱乐，消遣

text /tekst/   *v.* 发短信

irregular /ɪˈregjələ/   *adj.* 不规则的,不对称的；无规律的

summarize /ˈsʌməraɪz/   *vt.* 总结,概述

tablet /ˈtæblət/   *n.* 药片；丸   便笺本；平板电脑

heighten /ˈhaɪtn/   *v.* (使)变高；(使)加强

supplemental /ˌsʌplɪˈmentl/   *adj.* 补足的,追加的

implement /ˈɪmplɪment/   *vt.* 实施,执行；使生效,实现；落实(政策)

personalize /ˈpɜːsənəlaɪz/   *vt.* 使……拟人化

acronym /ˈækrənɪm/   *n.* 首字母缩略词

navigate /ˈnævɪɡeɪt/   *v.* 导航,确定路线；航行,航海；小心翼翼地绕过(障碍)；驾驭,成

         功应对(困难处境)

emulate /ˈemjuleɪt/   *v.* 仿效,模仿；竞争

target /ˈtɑːɡɪt/   *n.* 目标；对象；靶子   *v.* 以……为攻击目标；瞄准,面向

query /ˈkwɪəri/   *n.* 问题；疑问；询问；问号   *vt.* 质疑,对… 表示疑问

         vi. 询问；表示怀疑

get off 下(车、马等);离开;发出;(使)入睡

pay attention to 注意

post on 把……张贴在……上;不断向某人提供有关(某事)的消息

incorporate into 使成为……的一部分;并入;划归;归并

operate on 给……做手术;产生作用

rote memorization 机械背诵

at one's disposal 供任意使用,可自行支配

associate with 与……交往,联系;交接

take on 承担;呈现;雇用;录用

 ## *Notes*

1. The Intense School

The Intense School is American Online Education Schools. It has provided efficient IT training and certification services to 45,000 professionals in information technology and information security worldwide for over 12 years. Intense School employs professional lecturers to teach students, and students can firmly master the required knowledge and skills by providing in-depth internship training and pre-test knowledge system.

2. YouTube

YouTube is a video website, early company located in San Bruno, California. It was registered on February 15, 2005, founded by Chinese-American Chen Shijun and others, for users to download, watch and share films or short films.

 ## *Exercise*

A. Answer the following questions according to the text.

1. According to the passage, what effect will technology have on our life?

2. What is computer assisted learning?

3. What is the great thing about computer assisted learning?

4. In what way can taking tests on the computer help students?

5. How can computer assisted learning help students with languages?

B. Choose the best answer to each of the following questions according to the text.

1. According to the passage, what might shock us?

　　A) Technology can become a distraction that takes students away from the learning process.

B) Some form of Computer Assisted Learning has been applied in classrooms since the 1960s.

C) The use of electronic tools has better illustrated a point during teaching process.

D) The use of electronic tools has heightened engagement among students.

2. According to the passage, computer assisted learning doesn't include _____.

A) online courses in colleges

B) homeschooling

C) distance learning

D) self-learning

3. According to the passage, what tools can NOT be involved in computer assisted learning?

A) Tablets.

B) DVD players.

C) Televisions.

D) Cameras.

4. Which is NOT one of the great things about computer assisted learning?

A) It can be implemented in every type of classroom.

B) It can help students take classes at home either on their own or to supplement their other learning.

C) It can promote the interaction between students and teachers.

D) It can lead to a more in-depth understanding of the knowledge being transmitted.

5. It can be inferred from the passage that _____.

A) Computer assisted learning has no shortcomings.

B) Computer assisted learning can help students learn languages.

C) Computer assisted learning have many advantages.

D) Computer assisted learning is the best way to learn languages.

 Text B

The Pros and Cons of Online Learning

If you're at a point in your life where you're considering continuing your education, you may wonder if online learning is the right path for you.

Taking an online course requires a **notable** investment of time, effort, and money, so it's important to feel confident about your decision before moving forward. While online learning works **incredibly** well for some people, it's not for everyone.

We recently sat down with MIT xPRO Senior Instructional Designer and Program Manager Luke Hobson to explore the **pros and cons** of online learning and what to look for in an online course. If you're waiting for a sign about whether or not to **enroll in** that course you've been **eyeing**, you just might find it here.

■**Pros of Online Learning**

First, let's take a look at the true value of online learning by examining some of the benefits:

1. **Flexibility**

Online learning's most significant advantage is its flexibility. It's the reason millions of adults have chosen to continue their education and pursue certificates and degrees.

Asynchronous courses allow learners to complete work at their own pace, **empowering** them to find the **optimal** time to consume the content and submit assignments.

Some people are more attentive, focused, and creative in the mornings compared to the evenings and vice versa. Whatever works best for the learners should be the priority of the learning experience.

2. Community

When Luke asks people about their main reason for enrolling in a course, a common answer is networking and **community**.

Learners **crave** finding **like-minded** individuals who are going through the same experiences and have the same questions. They want to find a place where they belong. Being in the company of others who understand what they're going through can help online learners who are looking for support and motivation during challenging times and times that are worth celebrating.

Some learners have created study groups and book clubs that have **carried on** far beyond the end of the course—it's amazing what can grow from a single post on a discussion board!

3. Latest information

"Speed is a **massive** benefit of online learning," and according to Luke, it often doesn't get the attention it deserves.

"When we say speed, we don't mean being quick with learning. We mean actual speed to market. There are so many new ideas evolving within technical spaces that it's impossible to keep courses the way they were originally designed for a long period of time."

Luke notes that a program on Additive Manufacturing, Virtual Reality and Augmented Reality, or Nanotechnology must be checked and **updated** frequently. More formal learning **modalities** have difficulty changing content at this rapid pace. But within the online space, it's expected that the course content will change as quickly as the world itself does.

■ **Cons of Online Learning**

Now that we've looked at some of the biggest pros of online learning, let's examine a few of the **drawbacks**:

1. Learning environment

While many learners thrive in an asynchronous learning environment, others struggle. Some learners prefer live lessons and an instructor they can connect with multiple times a week. They need these interactions to feel supported and to persist.

Most learners within the online space identify themselves as **self-directed** learners, meaning they can learn on their own with the right environment, guidance, materials, and assignments. Learners should know themselves first and understand their preferences when it comes to what kind of environment will help them thrive.

2. Repetition

One drawback of online courses is that the structure can be **repetitive**: do a reading, respond to two discussion posts, submit an essay, repeat. After a while, some learners may feel **disengaged** from the learning experience.

There are online courses that break the **mold** and offer multiple kinds of learning activities, assessments, and content to make the learning experience come alive, but it may take some research to find them! Luke and his colleagues at MIT xPRO are **mindful** of designing courses that genuinely engage learners from beginning to end.

3. **Underestimation**

Luke has noticed that some learners underestimate how much work is required in an online course. They may mistakenly believe that online learning is somehow "easier" compared to in-person learning.

For those learners who **miscalculate** how long they will need to spend online or how challenging the assignments can be, changing that **mindset** is a difficult process. It's essential to **set aside** the right amount of time per week to **contribute to** the content, activities, and assignments. Creating personal deadlines and building a study routine are two best practices that successful online learners follow to hold themselves **accountable**.

■**Experience the Value of Online Learning: What to Look For in an Online Course**

You've probably gathered by now that not all online courses are created equal. On one end of the **spectrum**, there are methods of online learning that leave learners stunned by what a great experience they had. On the other end of the spectrum, some online learning courses are so disappointing that learners regret their decision to enroll.

If you want to experience the value of online learning, it's essential to pick the right course. Here's a quick list of what to look for:

- Feedback and connection to peers within the course platform.

  Interacting regularly with other learners makes a big difference. Luke and the MIT xPRO team use peer-reviewed feedback to give learners the opportunity to engage with each other's work.

- Proof of hard work.

  In the online learning space, proof of hard work often comes in the form of Continuing Education Units (CEUs) or specific **certifications**. MIT xPRO course participants who successfully complete one or more courses are **eligible** to receive CEUs, which many employers, licensing agencies, and professional associations accept as evidence of a participant's serious commitment to their professional development.

Online learning isn't for everyone, but with the right approach, it can be a valuable experience for many people.

(990 words)

【Words, Phrases and Expressions】

notable /ˈnəʊtəbl/　adj. 值得注意的；显著的；著名的　　n. 名人

incredibly /ɪnˈkredəbli/　adv. 难以置信地；很，极为

eye /aɪ/　n. 眼睛；视力；眼状物 vt. 定睛地看；注视；审视；细看

flexibility /ˌfleksəˈbɪlɪtɪ/　n. 柔韧性，机动性，灵活性；伸缩性；可塑度；柔度

asynchronous /eɪˈsɪŋkrənəs/　adj. 异步的

empower /ɪmˈpaʊə/　vt. 授权；准许；使能够；使控制局势

optimal /ˈɒptɪməl/　adj. 最佳的，最优的；最理想的

community /kəˈmjuːnəti/　n. 社区；社会团体；共同体；[生态]群落

crave /kreɪv/　v. 渴望，热望；恳求，恳请；要求，需要

like-minded /laɪk ˈmaɪndɪd/　adj. 具有相似意向或目的的；志趣相投的

massive /ˈmæsɪv/　adj. 大的，重的；大块的，大量的；结实的；大规模的

update /ˌʌpˈdeɪt/　vt. 更新，使现代化；校正，修正 n. 现代化；更新的信息；更新的行
为或事例

modality /məʊˈdæləti/　n. 形式；模式；方式

drawback /ˈdrɔːbæks/　n. 缺点，不利条件，障碍

self-directed /selfdɪˈrektɪd/　adj. 自我指导的

repetitive /rɪˈpetətɪv/　adj. 重复的，啰嗦的；复唱的

disengaged /ˌdɪsɪnˈgeɪdʒd/　adj. 自由的；闲散的；未约束的；脱离的
v. 分开（disengage 的过去式和过去分词）；解开；释放；使
（部队）脱离战斗

mold /məʊld/　n. 模子；模式；类型；霉 vt. 塑造；浇铸；用模子做；用泥土覆盖
vi. 对产生影响，形成；发霉

mindful /ˈmaɪndfl/　adj. 留心的，注意的；记住的，不忘的；警觉的；经意

underestimation /ˌʌndərˌestɪˈmeɪʃn/　n. 过低之估价；过低估计

miscalculate /ˌmɪsˈkælkjuleɪt/　v. 误算；对判断错误

mindset /ˈmaɪndset/　n. 观念模式，思维倾向，心态

accountable /əˈkaʊntəbl/　adj. 负有责任的，应对自己的行为做出说明的；可解释的；有
解释义务的；可说明的

spectrum /ˈspektrəm/　n. 光谱；波谱；范围；系列

certification /ˌsɜːtɪfɪˈkeɪʃn/　n. 证明，鉴定，证书

eligible /ˈelɪdʒəbl/　adj. 合适的；在（法律上或道德上）合格的；有资格当选的；称心如意的
n. 合格者；合适者；称心如意的人；合乎条件的人（或东西）

pros and cons　利弊；正反两方面；优缺点；赞成和反对的理由

enroll in   (使)加入

carry on   继续；接着；坚持；举行；进行；开展

set aside   把放置一旁;不理会;取消;留出

contribute to 捐献;促成;投稿;有助于

 Notes

1. MIT xPRO

MIT xPRO is the official learning platform of MIT. Its online learning programs leverage vetted content from world — renowned experts to make learning accessible anytime, anywhere. Designed using cutting — edge research in the neuroscience of learning, MIT xPRO programs are application focused, helping professionals build their skills on the job.

 Exercise

A. Answer the following questions according to the text.

1. According to the passage, what does taking an online course require?

2. Who do self-directed learners refer to , according to the passage?

3. What may some learners feel if the structure of the online courses is repetitive?

4. What are the two best practices that successful online learners should follow?

5. What did Luke and the MIT xPRO team use to help learners to engage with each other's work?

B. Read the article and decide whether the following statements are TRUE(T) or FALSE(F).

(   )1. Online learning works incredibly well for everyone.

(   )2. The main reason for people to enroll in a course is flexibility and community.

(   )3. For online courses, people expect that the course content will change as quickly as the world itself does.

(   )4. Some learners overestimate how much work is required in an online course.

(   )5. If people work hark, they may be rewarded with Continuing Education Units (CEUs) or specific certifications.

Part 3 Extending Skill: Referring to Other People's Ideas

You often need to talk about the ideas of other people in a lecture, a tutorial or a thesis. Referring to other people's ideas in your work can be done in three ways. They are:

- Summarizing or concising expression of the original work,
- Paraphrasing or restating a text or passage in other words and
- Quoting directly.

While using or referring to the work of some other people, there are two fundamental principles that should be kept in mind: firstly, you should clearly mention the source of the work from where you are referring; secondly, there should be a clear distinction between your original work and the references. It is your duty to clarify the user that the references are not his own work. Also the sources of such reference should be easily found. Now let's explain the three ways in detail:

■Summarizing

Summarizing involves putting the main idea(s) into your own words, including only the main point(s). Once again, it is necessary to attribute summarized ideas to the original source. Summaries are significantly shorter than the original and take a broad overview of the source material.

While summarizing a main point, the argument or the main idea of an author, just reference to the original work collectively, without page numbers is sufficient. If a general or well known work in a subject is summarized, the mention of resources is not required. But if you are referring to some numerical data or figures like population of a city or literacy rate of a country, mentioning the source of such data is very important so that the reader can verify the authenticity of such data from the original source. For example:

- **Daniel Solove gives a good description** of the issues around this in his 2004 book on technology and privacy in the information age.
- **Briefly, in his chapter on Information Privacy Law, he explains** how the exemptions and loopholes in the Act meant that it did not fully address the concerns which had led to it being passed.

■Paraphrasing

Paraphrasing means referring to other people's ideas in words of your own, emphasizing the portion of work which is most significant to your subject. While paraphrasing other people's ideas, you must be careful to choose the right words to express the same meanings and thoughts that the author wanted to convey. The idea that is conveyed to your readers must be the same, if the reader would have read the original work.

Paraphrasing must be done by using words entirely of your own. Just changing a few words from the original work doesn't account to paraphrasing. If you just change a few words of the paragraphs, it will caused plagiarism. While paraphrasing it is obligatory to mention the original source of work, typically with a page number. For example,

- **The original sentence:** When ICT first appeared in the workplace, many people feared they would lose their jobs to machines.
- **Paraphrase of the sentence:** The introduction of ICT in the workplace caused many employees to fear that they would be replaced by machines.

■ Quoting Directly

Quoting can be defined as putting the exact words or phrases of a person within quotation marks in your work. Inexperienced writers usually use quotations directly in their work. Excessive use of quotation within this is not considered as a good practice. Quotations should be sparingly used, where expressing the idea in your own words becomes very difficult or the phrase is a well known or generally accepted phrase in the subject. For example:

- "To support large-scale remote work, the platform tapped Alibaba Cloud to deploy more than 100,000 new cloud servers in just two hours last month — setting a new record for rapid capacity expansion," **according to DingTalk CEO, Chen Hang**.

【Exercise】

Read the following sentences and try to refer to them by summarizing or paraphrasing.

1. According to Luke, "When we say speed, we don't mean being quick with learning. We mean actual speed to market. There are so many new ideas evolving within technical spaces that it's impossible to keep courses the way they were originally designed for a long period of time." (Paraphrasing)

2. Learners crave finding like—minded individuals who are going through the same experiences and have the same questions. They want to find a place where they belong. Being in the company of others who understand what they're going through can help online learners who are looking for support and motivation during challenging times and times that are worth celebrating. (Paraphrasing)

3. Luke notes that a program on Additive Manufacturing, Virtual Reality and Augmented Reality, or Nanotechnology must be checked and updated frequently. More formal learning modalities have difficulty changing content at this rapid pace. But within the online space, it's expected that the course content will change as quickly as the world itself does. (Summarizing)

Part 4　Optional Exercises

A. Translate the following paragraph into Chinese.

Luke has noticed that some learners underestimate how much work is required in an online course. They may mistakenly believe that online learning is somehow "easier" compared to in－person learning. For those learners who miscalculate how long they will need to spend online or how challenging the assignments can be, changing that mindset is a difficult process. It's essential to set aside the right amount of time per week to contribute to the content, activities, and assignments. Creating personal deadlines and building a study routine are two best practices that successful online learners follow to hold themselves accountable.

B. Translate the following paragraph into English.

网上学习越来越受欢迎，因为它有很多好处。第一，网上学习是一种灵活的学习方式。传统方式的学习是坐在教室里面，但是现在，人们就算是坐在家里、咖啡店里，也能接触到知识。这是多么方便和高效啊！只要人们想学习就可以学到，不用担心位置的问题。第二，网上学习可以省很多钱。当人们想要参加传统方式的学习时，他们要交很多的钱，买书、请老师。然而网上学习可以省略这些不必要的麻烦，人们可以立刻听课，只要他们点击按钮，多么快速啊！网上学习是一种新的学习知识的途径，网上课程灵活、便宜、省时。有了这些优势，我相信将来网上学习会更受欢迎。

C. Write a composition on "Online Learning". Try to imagine what will happen when more and more people study online. You are required to write at least 120 words, but no more than 200 words.

Unit 7

Data Processing in Social Media

Part 1 Vocabulary

A. Choose the explanations in Column B that best match the terms in Column A.

Column A	Column B
1. social-media analytics	a. the awareness of market-level pricing intricacies and the impact on business, typically using modern data mining techniques
2. urban-sustainability	b. a technology-enabled discipline in which business and information technology work together to ensure the uniformity, accuracy, stewardship, semantic consistency and accountability of the enterprise's official shared master data assets
3. social-ecological-technological interactions	c. the process of gathering and analyzing data from social networks such as Facebook, Instagram, LinkedIn and Twitter.
4. Information and Communications Technology (ICT)	d. the practices to build them (urbanism), that focuses on promoting their long term viability by reducing consumption, waste and harmful impact
5. Business Intelligence	e. a three-phase process where data is first extracted then transformed (cleaned, sanitized, scrubbed) and finally loaded into an output data container
6. pricing intelligence	f. interactions between the considerations of the developments of the society and the protection of our environment
7. extract Transform Load	g. information that is delivered immediately after collection
8. master-data management (MDM)	h. the strategies and technologies used by enterprises for the data analysis and management of business information

Column A	Column B
9. real-time data	i. It is also used to refer to the convergence of audiovisual and telephone networks with computer networks through a single cabling or link system.
10. disaster risk reduction	j. a systematic approach to identifying, assessing and reducing the risks of disaster

B. Use the words from Exercise A to label the following pictures.

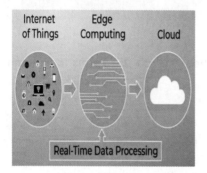

1. _____

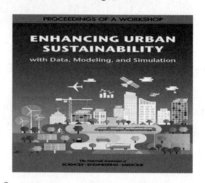

2. _____

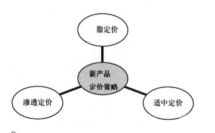

3. _____

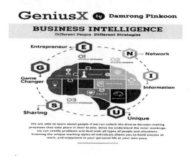

4. _____

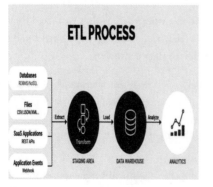

5. _____

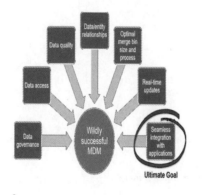

6. _____

C. Complete the following sentences with the words given below. Change the form if necessary.

voluminous	transient	futile	endeavor	ecological
prong	hamper	spatially	compromise	challenge
sustainability	interaction	expand	confine	associate

1. We do still need the human touch or the human _____, particularly when people are depressed.

2. He's been _____ to a wheelchair since the accident.

3. Police should have the power to fine people who _____ rescue efforts.

4. It would be _____ to sustain his life when there is no chance of any improvement.

5. The government has _____ with its critics over monetary policies.

6. The demonstrators have now made a direct _____ to the authority of the government.

7. When choosing what products to buy and which brands to buy from, more and more consumers are looking into _____.

8. A child's vocabulary _____ through reading.

9. We made an earnest _____ to persuade her.

10. Through science we've got the idea of _____ progress with the future.

Part 2   Reading

 Text A

Social-Media Data for Urban Sustainability

A **voluminous** and complex amount of information—'big data'—from social media such as Twitter and Flickr is now **ubiquitous** and of increasing interest to researchers studying human behaviour in cities. Yet the value of social-media data (SMD) for urban-**sustainability** research is still poorly understood. Here, we discuss key opportunities and challenges for the use of SMD by sustainability scholars in the natural and social sciences as well as by **practitioners** making daily decisions about urban systems. Evidence suggests that the vast scale and near-real-time observation are unique advantages of SMD and that solutions to most SMD challenges already exist.

We live on an urban planet. Human behaviour and values in cities are affecting, and may even drive, the future of global sustainability. As societies have become more globalized, dynamic and **transient**, planning for the sustainable city of tomorrow has become an elusive, and some argue even **futile endeavour**. Global urban science remains **fragmented** and disconnected from global and local policy and planning, highlighting the need for new tools and data to advance understanding of complex urban dynamics, and to support decision-making for sustainability transformations. A key question then is what new **heuristics** and data sources can help us capture the growing complexity of social-**ecological**-technological interactions in cities to help build a new global urban science and provide new knowledge for improving decision-making towards more-sustainable urban futures.

In the era of information and communications technology (ICT), the Internet of Things (IOT), many types of big data, and ubiquitous technology at our fingertips, urban **geolocated** data from social media promises to expand our understanding not only of where people are and what they do, but also what they value human behaviour and values are critical to **align** sustainability planning and policy with the needs and interests of residents. Near-real-time dynamic observations on unprecedented scales—city, regional and global—are now within reach. While research using data from social media such as Twitter, Instagram and Flickr has steadily grown in different scientific domains—such as digital humanities, urban ecology, **epidemiology**, tourism, disaster management and marketing—comprehensive accounts on the use of geolocated big data from social media for sustainable city planning are still rare. To our knowledge, to date no paper has reviewed the literature to **synthesize** the opportunities and challenges that SMD pose to sustainability research and provide examples of how scientists and practitioners are using SMD to advance sustainability goals. This is a **consequential** omission since more-livable, healthy and prosperous human settlements in the twenty-first century cannot be attained with the planning tools of the twentieth-century city. Here, we review evidence of SMD-based research carried out over the past decade, and discuss SMD as a critical component of big data and its applications for different spheres of sustainable urban development. We consider sustainable development to be grounded in the three-**pronged** approach suggested in the Brundtland Report but **take into account** more recent **iterations** put forward in the 2030 Agenda for Sustainable Development. We identify distinct advantages of SMD over traditional research methods and explicitly consider how SMD can aid the pursuit of environmental, public health, social equity, **infrastructural** and economic goals in cities. Next to untapped opportunities, the analysis reveals crosscutting challenges to the usability and reliability of SMD in scientific research. We demonstrate that although constraints such as missing **demographics**, **exiguous** locative data, **volatility** of social-media behaviour, and privacy concerns are pervasive, they are not **unsurmountable**. However, the opportunity to **take advantage of** new and massive data streams on human behaviour and perceptions is exciting. **Ancillary to** the growing volumes of publicly available urban data, SMD can offer city planners and engineers around the globe the opportunity to fill topical gaps and overcome the time lags affecting traditional census sources. Needs assessments and future-oriented planning can use

SMD as a starting point for established forms of community engagement and participatory GIS science. Because the pursuit of economic, environmental and social goals entails multiple trade-offs and compromises, cities can **leverage** SMD tools to facilitate discussion about how weights and values are to **be assigned to** different aspects of city life (for example, affordable housing versus open space). Additionally, if SMD are to help cities to build a much-needed global urban science, SMD-literate public officials and new spaces for big-data governance will have to be established.

There is much work remaining to overcome the existing challenges that face SMD and make the sector more **amenable** to the wide variety of research needs. Our Review suggests that the use of SMD for sustainability research is still confined to a subset of sustainability domains. Food insecurity, clean energy, quality education and gender equality are among the spheres where the contribution of SMD is yet to be fully tested. Culture, the fourth pillar of sustainability, has also received scant attention by SMD researchers.

Understanding the value of urban green space, as well as the use and social benefits associated with urban parks, is one additional area where SMD could provide new and important opportunities. Researchers have struggled for the last decade or more to bring values of nature more fully into planning, policymaking, and design and development opportunities. Despite the rise of urban tree planting and slow but steady progress integrating green space into urban infrastructure, fully accounting for the monetary value of urban green space for public physical and mental health, recreation, spiritual and aesthetic value, climate-change adaptation, disaster risk reduction and more has been hampered by lack of temporally and **spatially** comprehensive data on people's use of green space.

(918 words)

【Words, Phrases and Expressions】

voluminous /vəˈluːmɪnəs/　*adj.* 庞大的；很大的

ubiquitous /juːˈbɪkwɪtəs/　*adj.* 无处不在的；似乎无所不在的；十分普遍的

sustainability /səˌsteɪnəˈbɪlɪti/　*n.* 持续性

practitioner /prækˈtɪʃənə/　*n.* 专业人员；（尤指医学或法律界的）从业人员；习艺者；专门人才

transient /ˈtrænziənt/　*adj.* 转瞬即逝的；短暂的；临时的

futile /ˈfjuːtaɪl/　*adj.* 徒然的；徒劳的；无效的

endeavor /ɪnˈdɛvər/　*n./v.* 努力；尽力；试图

fragment /ˈfræɡˈment/　*v.* （使）碎裂，破裂，分裂

heuristics /hjuˈrɪstɪks/　*n.* 探索法；启发式

ecological /ˌiːkəˈlɒdʒɪkl/　*adj.* 生态的；关注生态环境的；主张生态保护的

geolocate /ˈdʒiəʊ ˈləʊkeɪ/　*v.* 地理定位功能

align /əˈlaɪn/　*v.* 排列；校准；排整齐；（尤指）使成一条直线；使一致

epidemiology /ˌepɪˌdiːmɪˈɒlədʒi/　*n.* 流行病学

synthesize /ˈsɪnθəsaɪz/   vt. (通过化学手段或生物过程)合成;综合

consequential /ˌkɒnsɪˈkwenʃl/   adj. 相应的;重要的;随之而来的;作为结果的

pronged /prɒŋd/   adj. 齐下;分为不同方向的;尖端分叉的

iteration /ˌɪtəˈreɪʃn/   adj. 迭代;(计算机)新版软件

infrastructural /ˌɪnfrəˈstrʌktʃərəl/   adj. 基础设施的

demographics /ˌdeməˈɡræfɪks/   n. 人口统计数据

exiguous /eɡˈzɪɡjuəs/   adj. 微小的;稀少的;不够的

volatility /ˌvɒləˈtɪlɪti/   n. 不稳定性,暂时性;挥发性(度)

unsurmountable /ˌʌnsərˈmaʊntəbl/   adj. 不能克服的,不能超越的

leverage /ˈliːvərɪdʒ/   n. 影响力

amenable /əˈmiːnəbl/   adj. 易控制的;顺从的;顺服的

spatially /ˈspeɪʃəli/   adv. 空间地;空间上;空间

take into account   考虑到;把……计算在内

take advantage of   利用;占……的便宜;欺骗;捉弄

ancillary to   增加;辅助

be assigned to   被指派;被分配给……;归属于

 # *Notes*

1. Brundtland Report

The Brundtland Report is a report which advocates the establishment of a long-term environmental strategy to map out viable development achievements suitable for the post-2000 period. It is recommended that collaborative mechanisms among different countries be developed to promote environmental improvement.

2. GIS

GIS is short for Geographic Information system, which is sometimes referred to as "Geo-Information system". It is a special and very important spatial information system. It is a technical system supported by computer hardware and software systems to collect, store, manage, calculate, analyze, display and describe the geographic distribution data in the whole or part of the earth's surface (including the atmosphere) space.

Exercise

A. Answer the following questions according to the text.

1. What is meant by "big data" according to the text?
2. What can people do to plan for the sustainable city of tomorrow?
3. What can we benefit from the growing volumes of publicly available urban data?
4. Why SMD can be used as a starting point for established forms of community engagement and participatory GIS science?
5. Which area can benefit from SMD that will provide new and important opportunities?

B. Choose the best answer to each of the following questions according to the text.

1. Where can we get the so-called "big-data"?
   A) From the library.   B) From the social media.
   C) From the teachers.   D) From the universities.
2. What are not critical to align sustainability planning and policy with the needs and interests of residents?
   A) where people are   B) what people do
   C) what people value   D) where people live
3. What problems do people find while reviewing SMD-based research carried out over the past decade?
   A) The use of geolocated big data from social media for sustainable city planning are still rare.
   B) The use of big-data for sustainable city planning is very promising.
   C) It is almost impossible to apple the data from social media to develop the cities.
   D) The use of the data from social media in education has been a big trend.
4. According to the passage, what other areas can SMD be fully tested?
   A) Food insecurity.   B) Clean energy.
   C) Gender equality.   D) Human behavior.
5. Which of the following about "urban green space" is false?
   A) The progress of urban tree planting has been steady and fast.
   B) Disaster risk reduction and more has been hampered by lack of green space.
   C) SMD could provide new and important opportunities for urban green space.
   D) Researchers have been dedicated to bring values of nature in urban green space.

Text B

Key Lessons for Converting Social Media Data into Business Intelligence

■ **Transform Real-Time Social Media Data into "Noise-Free" Actionable Intelligence**

Businesses that make more timely, accurate and informed decisions gain a significant competitive advantage. These decisions depend on relevant, real-time, quality data. However, gathering the data, **transforming** it into understandable

> **Social Media Data**
> · Twitter grew 1444% year/year
> · 50M tweets sent every day
> · Facebook has 400M+ active users
> · Every minute 600 new blog posts are published

information and loading it into business intelligence (BI) tools is easier said than done. And today, there's another issue at hand. How do you transform an **exponentially** expanding cloud of social media data into **actionable** intelligence?

The information exists. Data is refreshed in real-time, readily available as accurate economic indicators and detailed views of reality. The key is to figure out how to easily **extract** real-time intelligence before your competitors do, and to find clever ways to analyze these new sources.

■ **Timely, Accurate Information Improves Business Performance**

Fortunately, we are at a critical juncture where social media can be effectively **integrated into** enterprise Business Intelligence (BI) platforms. Real-time data offers a **compelling** alternative to traditional sources like government reports and labor department statistics that take months to compile. The technology is available and mature, and companies—including high profile brands and lower-profile innovators—are redefining their businesses by leveraging real-time social media data.

With more intelligent, more accurate real-time Web data processing, new methods for working with business information are possible. Business analysts and decision makers can then spend their time extracting greater intelligence from the data and less time worrying about collecting or accessing the data.

Think about the impact to your business if you could automatically add high-value Web data to your market intelligence, pricing intelligence, financial intelligence or any other business intelligence application. Until recently, this seemed like an impossible feat, or at least cost **prohibitive** based on the man hours involved.

■ **Traditional BI Processes are not Built for Agile Data Access**

While many businesses use BI **dashboards** to measure performance and comb through analytics data, a key piece of the overall BI strategy often goes missing. BI tools do an adequate job of analyzing data, but data quality is less of a priority overall. What's needed is BI functionality that delivers **agile** data access and high quality underlying data.

Typically, businesses employ expensive implementation cycles for gathering back-office data and feeding it into BI systems. Data moves **back and forth** between systems via tedious Extract Transform Load (ETL) methods. Complex master-data management (MDM) schemes are employed to make sense of everything. This process taxes internal IT departments, adds layers of expense, and consumes a lot of time. Other data comes from reports and Excel spreadsheets maintained by individual executives, and from "**siloed**" applications that require IT **intervention** to extract the information. When employees use "off-line" data for their reports, data quality suffers and synchronization with enterprise data becomes problematic. A better approach would be to leverage data where it resides. More than 5 billion Web sites crank out a vast amount of useful information. To quantify this in physical terms, a study found that the amount of digital information that will be created in 2010 alone will be 1. 2 zettabytes, which equals 75 billion fully-loaded 16GB Apple iPads! This **astronomical** growth is fueled by social media data.

This relevant, timely data is available, but it's not going to have an API any time soon—if ever. Data feeds from blogs, forums, Facebook, twitter and other social communities on the Web are predominantly unstructured text. This fountain of value contains some noise, and most enterprises view it as a fire hose. It's interesting but there's no way to consume it in an organized fashion and gain useful insights. Employing IT staff to make sense of the stream by developing custom APIs would be unrealistic at best and **foolhardy** at worst.

So, valuable information exists, it needs to be easily accessed, and structure needs to be applied to make it consumable by BI tools.

■Social Media—A New Frontier for Leading Edge Data Analysts

Economists, academics, governments and organizations of every kind have been using Web data services to automate access to accurate data and integrate it into analysis systems as soon as it's available. On April 8, 2010, *The Wall Street Journal* documented the trend in their article "New Ways to Read Economy-Experts Scour Oddball Data to Help See Trends Before Official Information is Available." The report provided numerous examples of how economists use unconventional data analytics to examine social and economic trends. They're analyzing everything from diesel fuel consumption to jobless claims.

The next logical step is to mine social media data. It's now easier for analysts and business decision makers to tap real-time data because a new, emerging class of ETL technologies takes the complexity out of the process, leaving more time for analysis and decision making (as opposed to information gathering).　　　　　(895 words)

【Words，Phrases and Expressions】

convert /kən'vəːrt/　　*v.* 转换；(使)转变；转化；可转变为；可变换成

transform /træns'fɔːrm/　　*vt.* 使改变；使改观；使转换

exponentially /ˌɪkspə'nənʃəli/　　*adv.* 以指数方式

actionable /'ækʃənəbl/　　*adj.* 能够使用的

extract /'ekstrækt/　　*vt.* 提取；设法得到；选取；选录；获得，得到

compelling /kəm'pelɪŋ/　　*adj.* 非常强烈的；不可抗拒的；令人信服的

prohibitive /prə'hɪbətɪv/　　*adj.* (以法令)禁止的；高昂得令人难以承受的

dashboard /'dæʃbɔːd/　　*n.* 仪表板

agile /'ædʒaɪl/　　*adj.* (动作)敏捷的；灵活的；(思维)机敏的；机灵的

siloed /saɪld/　　*adj.* 仓储式的

intervention /ˌɪntə'venʃn/　　*n.* 干涉；干预

astronomical /ˌæstrə'nɒmɪkl/　　*adj.* 极其巨大的

foolhardy /'fuːlhɑːdi/　　*adj.* 鲁莽的；有勇无谋的；莽撞的

integrated into　　融入；集成

back and forth　　来回地

Exercise

A. Answer the following questions according to the text.

1. How do you transform social media data into actionable intelligence?

2. How can the enterprise business benefit from real-time data?

3. What do the business analysts and decision makers do with more intelligent, more accurate real-time Web data processing?

4. Where can the data be collected on the web?

5. What can the report from *The Wall Street Journal* provide for enterprise business?

B. Read the following statements and decide whether they are TRUE(T) or FALSE(F).

(　　)1. It is easier for us to transform data into understandable information and loading it into business intelligence (BI) tools.

(　　)2. The effective processing of data will have great impact to your market intelligence, pricing intelligence, and financial intelligence.

(　　)3. Valuable information doesn't exist unless it can be applied into business intelligence.

(　　)4. Because of the ETL technologies, it is much easier for us to take the complexity out of the process.

(　　)5. The enterprises will redefine the businesses by leveraging real-time social media data.

Part 3　Extending Skill: Choosing the Correct Writing Plan

When you are given a written assignment, you must decide on the best writing plan before you begin to write the outline. Here are the insights for you to consult to help you choose the correct writing plan.

■**Descriptive Writing**

List the most important points of something: e. g. a list of key events in chronological order, history of a computer system; a description of a process, detailing each step and what the outcome of each step might be. Summarize points in a logical order.

Possible structure

a) Introduction

b) Description of process/system

c) Point/Step 1 and outcome

d) Point/Step 2 and outcome

e) Point/Step 3 and outcome

f) Conclusion

■**Analytical Writing**

List the important points which in your opinion explain the situation. Justify your opinion in each case. Look behind the facts at the how and why, not just what/who/when. Look for and question accepted ideas and assumptions.

Possible structure

a) Introduction

b) Definitions

c) Most important points

　　Example/Evidence/Reason 1

　　Example/Evidence/Reason 2

d) Other points

　　Example/Evidence/Reason 3

　　Example/Evidence/Reason 4

　　…

e) Conclusion

■**Comparison/Evaluation (may incorporate case studies)**

Decide on and define the aspects to compare two subjects. You may use these aspects as the basis for paragraphing. Evaluate which aspect(s) is(are) better or preferable and give reasons/criteria for your judgement.

Possible structure

a) Introduction

b) State and define aspects

   Either:

c) Aspect 1: Subject A v. s. B

d) Aspect 2: Subject C v. s. D

   Or:

c) Subject A: Aspect 1, 2, etc.

d) Subject B: Aspect 1,2, etc.

e) Conclusion/Evaluation

■**Argument Writing (may incorporate case studies)**

Analyze and/or evaluate your opinion, then write it down in a statement at the beginning or the end. Show awareness of difficulties and disagreements by mentioning counter-arguments. Support your opinion with evidence.

Possible structure 1

a)Introduction: Statement of issue

b) Statement of opinion

c) Define terms

d) Point 1: Explain and show the evidence

e) Point 2: Explain and show the evidence

f) Conclusion: Implications, etc.

Possible structure 2

a) Introduction: Statement of issue

b) Define terms

c) For: Point 1, 2, etc.

d) Against: Point 1, 2, etc

e) Conclusion: Statement of opinion

【Exercise】

Study the following writing assignments and identify which type of writing it belongs to.

1. You are supposed to write a composition on the topic **Jobs for Graduates** based on the outline below.

   ①大学生难找工作　　　　②原因很多　　　　③解决的办法

2. You are supposed to write a composition on the topic **The most unforgettable Person I ever Know** based on the outline below.

   ①我生活中最难忘的人是　　②为什么他(或她)令我难以忘怀　　③结论

3. You are supposed to write a composition on the topic **Pet Raising** based on the outline below.

①有些人喜欢将动物作为宠物　　②有些人反对将动物作为宠物圈养

4. You are supposed to write a composition on the topic **Work Is Necessity** and state the evidence or reasons.

Part 4　Optional Exercises

A. Translate the following paragraph into Chinese.

Fortunately, we are at a critical juncture where social media can be effectively integrated into enterprise Business Intelligence (BI) platforms. Real-time data offers a compelling alternative to traditional sources like government reports and labor department statistics that take months to compile. The technology is available and mature, and companies—including high profile brands and lower-profile innovators—are redefining their businesses by leveraging real-time social media data.

B. Translate the following paragraph into English.

社交媒体提高了生活信息的透明度。社会化媒体比任何其他技术都更能促进生活中的合作创新精神,从而使所有的公司都受到公众的监督。企业对社交媒体的热情越高,其透明度越高。在与社交媒体融合之前,大型企业很难与用户交互并获得反馈,在融入社交媒体后,用户可以与企业之间达到最高效的交流。此外,所有企业在面对环境问题、产品标准、消费者权益和员工聘用时更加谨慎。社交媒体提高了产品质量。社交媒体使所有消费者都能对产品进行评论和建议。因此,产品必须具有优异的质量。制造商生产的产品若不符合质量标准将被曝光并最终导致失败。这也是为什么好的产品在传统营销上花费较少的原因。社交媒体的存在使优秀的产品能够拥有自己的忠实用户和粉丝。

C. Write a composition on the topic "The Influence of Social-Networking on College Students". You should write at least 120 words.

Unit 8

Cloud Computing

Part 1   Vocabulary

A. Choose the explanations in Column B that best match the terms in Column A.

Column A	Column B
1. cluster	a. an architectural style that supports service orientation
2. grid computing	b. the virtualization/emulation of a computer system
3. Software as a Service(SaaS)	c. a commitment upon between a service provider and a client
4. High-Throughput Computing (HTC)	d. a system of running a computer program using a lot of small computers that are connected together in order to do very complicated jobs
5. Amazon Web Services(AWS)	e. a group of sectors on one or more computer disks
6. Virtual Machine(VM)	f. a software licensing and delivery model in which software is licensed on a subscription basis and is centrally hosted
7. Service-level Agreements(SLAs)	g. a nonparametric method in operations research and economics for the estimation of production frontiers
8. Data Envelopment Analysis (DEA)	h. the use of many computing resources over long periods of time to accomplish a computational task
9. Infrastructure as a Service (IaaS)	i. a cloud computing service model by means of which computing resources are hosted in a public, private, or hybrid cloud
10. Service-Oriented Architecture	j. a subsidiary of Amazon that provides on-demand cloud computing platforms and APIs to individuals, companies, and governments, on a metered pay-as-you-go basis

B. Use the words from Exercise A to label the following pictures.

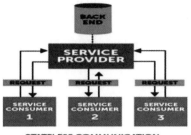

1. _____

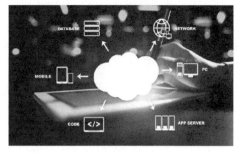

2. _____

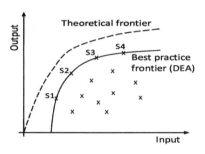

3. _____

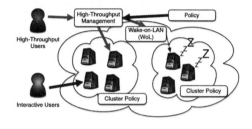

4. _____

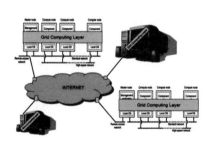

5. _____

6. _____

C. Complete the following sentences with the words given below. Change the form if necessary.

elastic	leverage	outsource	protocol	scenario
hybrid	customization	subscription	throughput	utilization
whereby	scale	provision	virtualization	portability

1. The new light cover increases this model's _____.

2. A _____ is a written record of a treaty or agreement that has been made by two or more countries.

3. His position as mayor gives him _____ to get things done.

4. The network can handle large _____.

5. The company began looking for ways to cut costs, which led to the decision

to _____.

6. They have introduced a new system _____ all employees must undergo regular training.

7. The worst-case _____ is an aircraft will crash if a bird destroys an engine.

8. Beat it until the dough is slightly _____.

9. _____ cars can go almost 600 miles between refueling.

10. You can become a member by paying the yearly _____.

Part 2   Reading

 Text A

Cloud Computing

■ Introduction

Cloud computing has been defined differently by many users and designers. IBM, a major player in cloud computing, has defined it as follows: " A cloud is a pool of **virtualized** computer resources. A cloud can host a variety of different **workloads**, including batch-style **back-end** jobs and interactive and user-facing applications. " The concept of cloud computing has evolved from cluster, **grid**, and utility computing. Cluster and grid computing leverage the use of many computers in parallel to solve problems of any size. Utility and Software as a Service (SaaS) provide computing resources as service with the notion of pay per use. Cloud computing leverages dynamic resources to deliver large numbers of services to end users. Cloud computing is a high-**throughput** computing (HTC) paradigm **whereby** the infrastructure provides the services through a large data center or server farms. The cloud computing model enables users to share access to resources from anywhere at any time through their connected devices. The cloud will free users to focus on user application development and create business value by **outsourcing** job execution to cloud providers. In this **scenario**, the computations (programs) are sent to where the data is located, rather than copying the data to millions of **desktops** as in the traditional approach. Cloud computing avoids large data movement, resulting in much better network bandwidth **utilization**. Furthermore, machine **virtualization** has enhanced resource utilization, increased application flexibility, and reduced the total cost of using virtualized data-center. The cloud offers significant benefit to IT companies by freeing them from the low-level task of setting up the hardware (servers) and managing the system software. Cloud computing applies a virtual platform with **elastic** resources put together by on-demand **provisioning** of hardware, software, and data sets, dynamically. The main idea is to move desktop computing to a service-oriented platform using server clusters and huge databases at data centers. Cloud computing leverages its low cost and simplicity to benefit both providers and users. Cloud computing intends to leverage multitasking achieve higher throughput by serving many heterogeneous applications, large or small, simultaneously.

■Public, Private, and Hybrid Clouds (Figure 8 – 1)

A public cloud is built over the Internet and can be accessed by any user who has paid for the service. Public clouds are owned by service providers and are accessible through a **subscription**. Many public clouds are available, including Google App Engine (GAE), Amazon Web Services(AWS), Microsoft Azure, IBM Blue Cloud, and Saleforce's Force.com. The providers of the aforementioned clouds are commercial providers that offer a publicly accessible remote interface for creating and managing VM instances within their proprietary infrastructure. A public cloud delivers a selected set of business processes. The application and infrastructure services are offered on a flexible price-per-use basis.

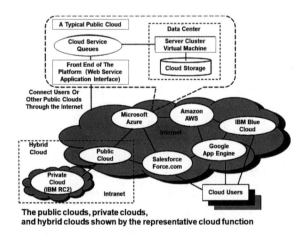

The public clouds, private clouds, and hybrid clouds shown by the representative cloud function

Figure 8 – 1

A private cloud is built within the domain of an Intranet owned by a single organization. Therefore, it is client owned and managed, and its access is limited to the owning clients and their partners. Its deployment was not meant to sell capacity over the Internet through publicly accessible interfaces. Private clouds give local users a flexible and agile private infrastructure to run service workloads within their administrative domains. A private cloud is supposed to deliver more efficient and convenient cloud services. It may impact the cloud standardization, while retaining greater customization and organizational control. Intranet-based private clouds are linked to public clouds to get additional resources.

A **hybrid** cloud is built with both public and private clouds, as shown at the lower-left corner of Figure 8 - 1. Private clouds can also support a hybrid cloud model by supplementing local infrastructure with computing capacity from an external public cloud. For example, the Research Compute Cloud (RC2) is a private cloud, built by IBM, that interconnects the computing and IT resources at eight IBM Research Centers scattered throughout the United States, Europe, and Asia.

A hybrid cloud provides access to clients, the partner network, and third parties. In summary, public clouds promote standardization, preserve capital investment, and offer application flexibility. Private clouds attempt to achieve **customization** and offer higher efficiency, **resiliency**, security, and privacy. Hybrid clouds operate in the middle, with many compromises in terms of resource sharing.

■**Cloud Design Objectives**

Despite the controversy surrounding the replacement of desktop or **deskside** computing by centralized computing and storage services at data centers or big IT companies, the cloud computing community has reached some consensus on what has to be done to make cloud computing universally acceptable. The are six design objectives for cloud computing:

Shifting computing from desktops to data centers: Computer processing, storage, and software delivery is shifted away from desktops and local servers and toward data centers over the Internet.

Service provisioning and cloud economics: Providers supply cloud services by signing service-level agreements (SLAs) with consumers and end users. The services must be efficient in terms of computing, storage, and power consumption. Pricing is based on a **pay-as-you-go** policy.

Scalability in performance: The cloud platforms and software and infrastructure services must be able to scale in performance as the number of users increases.

Data privacy protection: Can you trust data centers to handle your private data and records? This concern must be addressed to make clouds successful as trusted services.

High quality of cloud services: The Quality of Service (QoS) of cloud computing must be standardized to make clouds **interoperable** among multiple providers.

New standards and interfaces: This refers to solving the data **lock-in** problem associated with data centers or cloud providers. Universally accepted APIs and access **protocols** are needed to provide high **portability** and flexibility of virtualized applications.

(917 words)

【Words, Phrases and Expressions】

virtualize/ˈvɜːtʃʊəlaɪz/　*vt.* 虚拟化

workload /ˈwɜːkləʊd/　*n.* 工作量

back-end/ˈbækend/　*n.* 后端

grid /grɪd/　n. 网格；格子，栅格；输电网

throughput /ˈθruːpʊt/　n. 生产量；生产能力，吞吐量

whereby /weəˈbaɪ/　adv. 凭借；通过……；借以；与……一致

outsource /aʊtˈsɔːs/　vt. 把……外包　vi. 外包

scenario /sɪˈnɑːrɪəʊ/　n. 方案；情节；剧本

desktop /deskˌtɒp/　n. 桌面；台式机

utilization /ˌjuːtɪlaɪˈzeɪʃən/　n. 利用，使用

virtualization /ˈvɜːtʃʊəˌlaɪzeɪʃn/　n. 虚拟化

elastic /ɪˈlæstɪk/　n. 松紧带；橡皮圈　adj. 有弹性的；灵活的；易伸缩的

provision /prəˈvɪʒn/　n. 规定；条款；准备；供应品
　　　　　　　　　　　vt. 供给……食物及必需品

hybrid /ˈhaɪbrɪd/　n. 混合物　adj. 混合的

subscription /səbˈskrɪpʃn/　n. 捐献；订阅；订金

customization /ˌkʌstəmaɪˈzeɪʃən/　n. 定制；用户化；客制化服务

resiliency /rɪˈzɪlɪənsi/　n. 弹性；跳回；复原力

deskside /deskˌsaɪd/　adj. 桌边（型）的

pay-as-you-go /ˈpeɪəzjuˈɡəu/　n./adj. 现收现付制（的）；量入为出（的）；即付即用（的）

interoperable /ɪntərˈɒpərəbl/　adj. 彼此协作的；能共同操作的；能共同使用的

lock-in /ˈlɔkɪn/　n. 锁定；同步

protocol /ˈprəutəkɒl/　n. 礼仪，礼节；国际议定书，协议；条约草案，（协议或条约的）附
　　　　　　　　　　件；（协定，公约）修正案（或增补）

portability /ˌpɔːtəˈbɪləti/　n. 可移植性；轻便；可携带性

Exercise

A. Answer the following questions according to the text.

1. According to IBM, what's cloud computing?

2. What's the role played by machine virtualization in promoting utilization?

3. How do IT companies benefit from the cloud?

4. Can you list some public clouds which are available?

5. What are the design objectives for cloud computing?

B. Choose the best answer to each of the following questions according to the text.

1. According to this passage, how does cloud computing contribute to better network bandwidth utilization?

　A) Through hosting different workloads.

　B) By leveraging the use of many computers.

C) Via large numbers of services.

D) By evading large data movement.

2. Why is cloud computing regarded as a high-throughput computing (HTC) paradigm?

A) The services are provided by the infrastructure through a large data center.

B) Evolving from cluster, grid, and utility computing.

C) Providing computing resources as service.

D) Using dynamic resources.

3. The following elements are crucial in the application of a virtual platform for cloud computing except _____?

A) big data   B) hardware   C) software   D) data sets

4. How does cloud computing utilize multitasking to achieve higher throughput?

A) By centralized computing and storage services at data centers or big IT companies.

B) By serving a battery of different applications, large or small, simultaneously.

C) Via supplementing local infrastructure with computing capacity.

D) Through signing service-level agreements.

5. Which of the following is not the feature of public clouds?

A) promote standardization   B) preserve capital investment

C) provide security   D) offer application flexibility

 Text B

Look-Ahead Energy Efficient VM Allocation Approach for Data Centers

Energy efficiency is an important issue for reducing environmental **dissipation**. Energy efficient resource provisioning in cloud environments is a challenging problem because of its dynamic nature and varied application workload characteristics. In the literature, live migration of virtual machines (VMs) among servers is commonly proposed to reduce energy consumption and to **optimize** resource usage, although it comes with essential drawbacks, such as migration cost and performance **degradation**. Energy efficient provisioning is addressed at the data center level in this research. A novel efficient resource management algorithm for virtualized data centers that optimizes the number of servers to meet the requirements of dynamic workloads without migration is proposed in this paper. The proposed approach, named Look-ahead Energy Efficient VM Allocation (LAA), contains a Holt Winters-based prediction module. Energy efficiency and performance are **inversely** proportional. The energy-performance trade-off relies on periodic comparisons of the predicted and active numbers of servers. To evaluate the proposed algorithm, experiments are conducted with real-world workload traces from Google Cluster. LAA is compared with the best approach provided by CloudSim based on VM migration called Local Regression-Minimum Migration Time (LR-MMT). The experimental results show that the proposed algorithm leads to a consumption reduction of up to 45% to complete one workload compared with the LR-MMT.

Cloud computing is a collection of computer system resources that are dynamically provisioned to provide services to users based on demand access. Service providers offer customers three services, Software as a Service (SaaS), Platform as a Service (PaaS), and Infrastructure as a Service (IaaS), through data centers. As customers' needs for the services offered by data centers increase, the amount of energy consumed by data centers increases **linearly**. Increases of 48% and 34% are estimated for total world energy consumption and CO_2 emissions, respectively, between 2010 and 2040. Service providers try to reduce the energy cost in data centers due to laws, regulations, and standards. In addition, reducing the cost of the services and increasing the profit rate are other goals of providers. On the other hand, users also want to have the same service with acceptable quality and less cost which are defined through Service Level Agreement (SLA). For this reason, while aiming to reduce energy consumption and cost, the performance of the service offered to users should also be considered. However, it is not easy to address the needs of the users and the resources that will meet these requirements. Improving the energy efficiency of data centers has received significant attention in recent years. When cloud data centers are running at low usage levels of computing capacity without optimization, it causes high energy inefficiency. Many existing studies have proposed running servers' computational units at full capacity to increase energy efficiency, but this causes performance degradation. To solve performance degradation caused by running at full capacity, the static optimal utilization **threshold** is defined for each resource type, including CPU, RAM, bandwidth and so on. However, a **static** threshold may lead to machines being turned on or off unnecessarily since the resource demand in the future is not considered. Another approach in the literature is server consolidation, which reduces the number of active physical machines through VM migration or by collocating VMs to a small set of physical machines. However, VM migration and server consolidation techniques cause low throughput from the perspective of the service consumer as well as energy **overheads** from the perspective of the service provider. To address performance and cost issues, we propose an energy efficient resource allocation approach that integrates the Holt Winters **forecasting** model for optimizing energy consumption. The proposed approach includes a forecasting module to take into account not only the current situation but also the potential customer resource needs in the future. The approach is based on an adaptive decision mechanism for turning servers on/off and detecting under/over utilization. This approach is designed to avoid performance degradation and improve energy efficiency. The basis of this approach is based on our previous work. The forecasting module is added as promised in previous paper. Results have been analyzed more **robust** through sensitivity analysis and preferred comparison methods. Moreover, google cluster data are used to evaluate the algorithm with a real-world workload. The algorithm is implemented and run on CloudSim, which is a commonly used Cloud simulator as in the previous study. CloudSim provides several VM allocation and migration policies and is mainly focused on IaaS-related operations. However, the provided allocation algorithms run with static workloads. The existing code of CloudSim is extended to meet dynamic workload requirements and makes a fair comparison with the proposed algorithm.

The key contributions of this research are as follows:

We propose an energy efficient resource allocation algorithm called Look-ahead Energy Efficient Resource Allocation (LAA). LAA **facilitates** adaptive allocation of the incoming user requests to computing resources. A single threshold is used for CPU over-utilization detection, but it is not the only **parameter** used to decide whether to turn a new server on to host the newly arrived workload. The trend of the system is as important as the threshold during the allocation decision process. If the number of already active servers meets the forecasted future requirement and the current over-utilization situation occurs **temporarily**, then the proposed algorithm makes the allocation decision for the newly arrived workload by considering the remaining time for already running workloads on active servers.

The proposed algorithm **minimizes** energy consumption while preventing performance degradation. The algorithm is based on not only the current state of the system but also future demand as determined through Holt Winters forecasting to make adaptive decisions. Google has published the trace data of their clusters. These data are used to evaluate the performance of the proposed algorithm. After data analysis and comparison with another time series analysis **methodology** called Auto Regressive Integrated Moving Average (ARIMA), Holt Winters gave the better result in terms of the minimum error rate.

(996 words)

【Words, Phrases and Expressions】

dissipation /ˌdɪsɪˈpeɪʃn/   *n.* 浪费；消散；损耗

optimize /ˈɒptɪmaɪz/   *v.* 优化，充分利用（形势，机会，资源）

degradation /ˌdegrəˈdeɪʃn/   *n.* 损害；恶化；衰退

inversely /ˌɪnˈvɜːsli/   *adv.* 相反地；倒转地

linearly /ˈlɪniəli/   *adv.* 成直线地；在线上地

threshold /ˈθreʃhəʊld/ *n.* 门槛；门口

static /ˈstætɪk/   *adj.* 静止的，停滞的；静电的；静力的

overhead /ˌəʊvəˈhed/   *adj.* 经费的；管理费用的

forecast /ˈfɔːkɑːst/   *v.* 预测；预报

robust /rəʊˈbʌst/   *adj.* 强健的；强壮的

facilitate /fəˈsɪlɪteɪt/   *v.* 使更容易；使便利

parameter /pəˈræmɪtə/   *n.* 界限，范围；参数，变量

temporarily /ˈtemprərəli/   *adv.* 暂时地，临时地

minimize /ˈmɪnɪmaɪz/   *v.* 使减少到最低限度；贬低；使显得不重要

methodology /ˌmeθəˈdɒlədʒi/   *n.* 方法论；一套方法

 Exercise

A. Answer the following questions according to the text.

1. Why is energy efficient resource provisioning in cloud environments a challenging problem?

2. What are the demerits of live migration of virtual machines (VMs) among servers?

3. How many services do service providers offer?

4. What drives service providers to reduce energy cost in data centers?

5. According to the passage, what is CloudSim?

B. Read the following statements and decide whether they are TRUE(T) or FALSE(F).

(   )1. When cloud data centers are running at low usage levels of computing capacity without optimization, it makes no difference on energy efficiency.

(   )2. Energy efficiency and performance are directly proportional.

(   )3. In order to assess the proposed algorithm, experiments are conducted with virtual workload traces from Google Cluster.

(   )4. Averting performance degradation and enhancing energy efficiency are the main purposes of the LAA approach.

(   )5. The LAA approach adopted sensitivity analysis and comparison methods to analyze the results.

Part 3   Extending Skill: Presentation

If you want to deliver an effective presentation, PowerPoint will help you to engage the audience's attention and enhance what you are saying. The following are some steps of preparing a PowerPoint.

■To Brainstorm

- To brainstorm means to discuss ideas or ways to solve a problem.
- Before you create a PowerPoint, you should brainstorm its main points.
- You can ask yourself some questions like:
  What are the main points that I want to present?
  What information will my audience be interested to know?
- We brainstormed some ideas to make our presentation more engaging.

■To Outline

- To outline something means to roughly plan out the points or parts of it.
- Once you're brainstormed some ideas, you can outline your PowerPoint.
- This will help you structure your presentation.

■To Structure

- To structure something means to organize or arrange it in a particular way.
- After you've outlined your PowerPoint, you need to structure it.
- Structuring your PowerPoint allows you to list your main points clearly and logically.

■To Design

- To design something means to decide how it will look.
- After you have structured your PowerPoint, you can begin to design it.
- When designing a PowerPoint, you should consider its layout, visuals, and font.

■To Revise

- To revise something means to change something to improve it.
- Once you have finished your PowerPoint, you may want to revise it.
- You may be able to make it even better before you present it.

The use of PowerPoint slides is also critical in delivering a presentation. Here are several tips for a PowerPoint presentation.

1. Use the outline view first

Use the outline view by clicking on the view menu and selecting the outline command to ensure the content of your presentation is solid before deciding on the visual elements such as background, fonts, and graphic designs.

2. Avoid too much text

Avoid too much text on your slides. Use bullet points on slides. Better to use simple language and limit the number of bullets to three or four per slide with no more than eight words per line.

3. Limit the number of your slides

Slides used in a presentation should be spare. On average, one slide per minute is about right.

4. Use fonts that are easy to read

Select big fonts so that the audience can read it. A 28-or 32-point size is preferable forthe text, with titles being 36-to 44-point size. However, the best way to check is to pay a visit to the back of the presentation room prior to the talk and try it out.

5. Use colors effectively

The colors of your text and graphics should be in sharp contrast with the background so that whatever is on the slides is readable to your audience.

6. Keep multimedia to a minimum

Resist the temptation to overuse movies and sound files or other multimedia objects, if you want to keep the audience focused on you.

【Exercise】

Based on the content of this unit, make a PowerPoint presentation. Employ the above strategy.

Part 4　Optional Exercises

A. Translate the following paragraph into Chinese.

Cloud computing is a collection of computer system resources that are dynamically provisioned to provide services to users based on demand access. Service providers offer customers three services, Software as a Service (SaaS), Platform as a Service (PaaS), and Infrastructure as a Service (IaaS), through data centers. As customers' needs for the services offered by data centers increase, the amount of energy consumed by data centers increases linearly. Increases of 48% and 34% are estimated for total world energy consumption and CO_2 emissions, respectively, between 2010 and 2040. Service providers try to reduce the energy cost in data centers due to laws, regulations, and standards. In addition, reducing the cost of the services and increasing the profit rate are other goals of providers. On the other hand, users also want to have the same service with acceptable quality and less cost which are defined through Service Level Agreement (SLA). For this reason, while aiming to reduce energy consumption and cost, the performance of the service offered to users should also be considered. However, it is not easy to address the needs of the users and the resources that will meet these requirements. Improving the energy efficiency of data centers has received significant attention in recent years.

B. Translate the following paragraph into English.

    我这三十年，只做一件事情，就是给我们的地震学家和气象学家设计最快的超级电脑。使他们能够用最快速的计算工具进行复杂的计算和做各种研究。简单地说，云计算是一种基于因特网的超级计算模式，它有感知、有互联、有协作，能够随时随地感知任何事物。比如，科学家可以将云计算用于地震、水灾、旱灾、沙尘暴等自然灾害的预警上。除了自然灾害以外，我们还能把云计算用于人为灾害的预警上，包括环境污染、金融风暴、资源短缺，甚至政变、战争，等等。云计算还可以带来教育互动，将教育普及到世界每个角落，使资源得到有效利用。云计算还可以进入智慧家庭，甚至影响我们下一代的社会价值观：比如用云计算带动他们从小开始做公益，帮助他人。我希望在云计算的新社会里，新的智慧家庭会有新的突破。

C. Write a composition on "How has the Internet changed the way people live". You may refer to the outline given below.

    a. As a primary source of information as well as an efficient means of communication, the Internet offers people a colourful and convenient life.

    b. However, many students spend a lot of time on it, and they have become Internet addicts.

    c. What should we do?

Unit 9

Artificial Intelligence

Part 1 Vocabulary

A. Choose the explanations in Column B that best match the terms in Column A.

Column A	Column B
1. computer vision	a. an advanced type of machine learning architecture employed by neural networks, most commonly by "convolutional neural networks"
2. deep learning	b. a procedure to test whether a computer is capable of human-like thought, which was proposed by the British mathematician Alan Turing in 1950
3. feature extraction	c. the process of analyzing an acoustic speech signal to identify the linguistic message that was intended, so that a machine can correctly respond to spoken commands
4. Turing Test	d. the use of digital computer techniques to extract, characterize and interpret information in visual images of a three-dimensional world
5. speech recognition	e. a machine with the ability to apply intelligence to any problem, rather than just one specific problem, sometimes considered to require consciousness, sentience and mind, contrasted with weak AI
6. strong AI	f. starting from an initial set of measured data and building derived values (features) intended to be informative and non-redundant, facilitating the subsequent learning and generalization steps

Column A	Column B
7. machine learning	g. the computer architecture modeled upon the human brain's interconnected system of neurons
8. autonomous vehicle	h. an automated method of biometric identification that uses mathematical pattern-recognition techniques on video images of the irides of an individual's eyes
9. iris recognition	i. the process or technique by which a device modifies its own behavior as the result of its past experience and performance
10. neural network	j. a kind of car or truck that is able to plan its path and to execute its plan without human intervention

B. Use the words given below to label the following pictures.

ENIGMA code	iris recognition	natural language processing
machine learning	Turing Test	convolutional neural network

1. _____

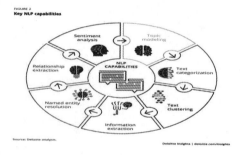

2. _____

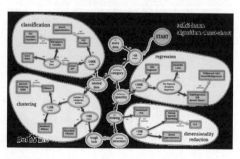

3. _____

4. _____

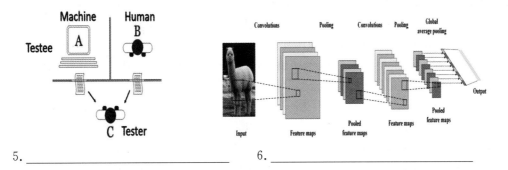

5. _____   6. _____

C. Complete the following sentences with appropriate words below. Change the form if necessary.

potential	rationality	confine	encompass	discipline
classification	application	consciousness	surpass	hierarchy
supervise	incorporate	derive	optimize	evolution

1. He's been _____ to a wheelchair since the accident.
2. Smartphone-based data collection comes at an appropriate time in the _____ of psychological science.
3. Unqualified members of staff at the hospital were not sufficiently _____.
4. Many people don't use their computers to their full _____.
5. When we acquire new information, the brain automatically tries to _____ it within existing information by forming associations.
6. Anna's strength is _____ from her parents and her sisters.
7. He was determined to _____ the achievements of his older brothers.
8. I can't remember any more — I must have lost _____.
9. Engineering and technology are _____ distinct from one another and from science.
10. Students learned the practical _____ of the theory they had learned in the classroom.

Part 2   Reading

 Text A

What is Artificial Intelligence (AI)

The birth of the artificial intelligence conversation was **denoted** by Alan Turing's

influential work, "Computing Machinery and Intelligence" (1950). In this paper, Turing, often referred to as the "father of computer science", asks the following question, "Can machines think?" From there, he offers a test, now famously known as the "Turing Test", where a human **interrogator** would try to distinguish between a computer and human text response.

Stuart Russell and Peter Norvig then proceeded to publish, *Artificial Intelligence: A Modern Approach* (1995), becoming one of the leading textbooks in the study of AI. In this book, they delve into four potential goals or definitions of AI, which **differentiates** computer systems on the basis of **rationality** and thinking vs. acting:

Human approach:

- Systems that think like humans
- Systems that act like humans

Ideal approach:

- Systems that think rationally
- Systems that act rationally

Alan Turing's definition would have fallen under the category of "systems that act like humans."

In 2004, John McCarthy offers the following definition in his paper, "What is artificial intelligence?", "It is the science and engineering of making intelligent machines, especially intelligent computer programs. It is related to the similar task of using computers to understand human intelligence, but AI does not have to **confine** itself **to** methods that are **biologically observable**."

At its simplest form, artificial intelligence is a field, which combines computer science and robust **datasets**, to enable problem-solving. It also encompasses **sub-fields** of machine learning and deep learning, which are frequently mentioned **in conjunction with** artificial intelligence. These disciplines are comprised of AI algorithms which seek to create expert systems which make **predictions** or **classifications** based on input data.

■**Types of Artificial Intelligence—Weak AI vs. Strong AI**

Weak AI—also called Narrow AI or Artificial Narrow Intelligence (ANI)—is AI trained and focused to perform specific tasks. Weak AI drives most of the AI that surrounds us today. "Narrow" might be a more accurate **descriptor** for this type of AI as it is anything but weak; it enables some very robust applications, such as Apple's Siri, Amazon's Alexa, IBM Watson, and autonomous vehicles.

Strong AI is made up of Artificial General Intelligence (AGI) and Artificial Super Intelligence (ASI). Artificial general intelligence (AGI), or general AI, is a theoretical form of AI where a machine would have an intelligence equaled to humans; it would have a self-aware **consciousness** that has the ability to solve problems, learn, and plan for the future. Artificial Super Intelligence (ASI)—also known as **superintelligence**—

would **surpass** the intelligence and ability of the human brain.

Deep learning vs. Machine learning

Both deep learning and machine learning are sub-fields of artificial intelligence, and deep learning is actually a sub-field of machine learning(Figure 9 – 1).

Deep learning is actually comprised of neural networks. "Deep" in deep learning refers to a neural network comprised of more than three layers—which would be inclusive of the inputs and the outputs—can be considered a deep learning algorithm. This is generally represented using the following diagram(Figure 9 – 2).

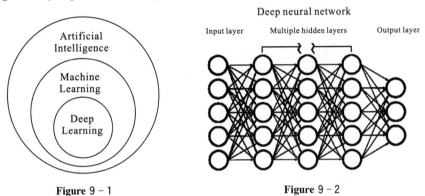

Figure 9 – 1   Figure 9 – 2

The way in which deep learning and machine learning differ is in how each algorithm learns. Deep learning **automates** much of the feature extraction piece of the process, eliminating some of the manual human intervention required and enabling the use of larger data sets. You can think of deep learning as "**scalable** machine learning". Classical, or "non-deep" machine learning is more dependent on human intervention to learn. Human experts determine the **hierarchy** of features to understand the differences between data inputs, usually requiring more structured data to learn.

"Deep" machine learning can leverage labeled datasets, also known as **supervised** learning, to inform its algorithm, but it doesn't necessarily require a labeled dataset. It can **ingest** unstructured data in its raw form (e. g. text, images), and it can automatically determine the hierarchy of features which distinguish different categories of data from one another. Unlike machine learning, it doesn't require human intervention to process data, allowing us to scale machine learning in more interesting ways.

■**Artificial Intelligence Applications**

There are numerous, real-world applications of AI systems today. Below are some of the most common examples.

· **Speech recognition:** It is also known as automatic speech recognition (ASR), computer speech recognition, or speech-to-text, and it is a capability which uses natural language processing (NLP) to process human speech into a written format. Many mobile devices **incorporate** speech recognition **into** their systems to conduct voice search (e.g. Siri) or provide more **accessibility** around texting.

- **Customer service:** Online virtual agents are replacing human agents along the customer journey. They answer frequently asked questions (FAQs) around topics, or provide personalized advice, cross-selling products or suggesting sizes for users, changing the way we think about customer engagement across websites and social media platforms. Examples include messaging bots on e-commerce sites with virtual agents, messaging apps, such as Slack and Facebook Messenger, and tasks usually done by virtual assistants and voice assistants.

- **Computer vision:** This AI technology enables computers and systems to **derive** meaningful information **from** digital images, videos and other visual inputs, and based on those inputs, it can take action. This ability to provide recommendations distinguishes it from image recognition tasks. Powered by convolutional neural networks(CNN), computer vision has applications within photo tagging in social media, radiology imaging in healthcare, and self-driving cars within the automotive industry.

- **Recommendation engines:** Using past consumption behavior data, AI algorithms can help to discover data trends that can be used to develop more effective cross-selling strategies. This is used to make relevant add-on recommendations to customers during the checkout process for online retailers.

- **Automated stock trading:** Designed to optimize stock **portfolios**, AI-driven high-frequency trading platforms make thousands or even millions of trades per day without human intervention.

■**History of Artificial Intelligence: Key Dates and Names**

The idea of "a machine that thinks" dates back to ancient Greece. But since the advent of electronic computing, important events and **milestones** in the **evolution** of artificial intelligence include the following:

- 1950: Alan Turing publishes "Computing Machinery and Intelligence". In the paper, Turing—famous for breaking the Nazi's ENIGMA code during World War Ⅱ—proposes to answer the question "Can machines think?" and introduces the Turing Test to determine if a computer can demonstrate the same intelligence (or the results of the same intelligence) as a human.

- 1956: John McCarthy **coins** the term "artificial intelligence" at the first-ever AI conference at Dartmouth College. Later that year, Allen Newell, J. C. Shaw, and Herbert Simon create the Logic Theorist, the first-ever running AI software program.

- 1967: Frank Rosenblatt builds the Mark 1 **Perceptron**, the first computer based on a neural network that "learned" through **trial and error**. Just a year later, Marvin Minsky and Seymour Papert publish a book titled *Perceptrons*, which becomes both the landmark work on neural networks and, at least for a while, an argument against future neural network research projects.

- 1980s: Neural networks which use a backpropagation algorithm to train itself become widely used in AI applications.
- 1997：IBM's Deep Blue beats the world chess champion Garry Kasparov, in a chess match (and rematch).
- 2011：IBM Watson beats champions Ken Jennings and Brad Rutter at *Jeopardy*!
- 2015：Baidu's Minwa supercomputer uses a special kind of deep neural network called a convolutional neural network to identify and categorize images with a higher rate of accuracy than the average human.
- 2016：DeepMind's AlphaGo program, powered by a deep neural network, beats Lee Sodol, the world champion Go player, in a five-game match.

(1266 words)

【Words, Phrases and Expressions】

denote /dɪ'nəʊt/　*vt.* 标志;预示;象征;表示

influential /ˌɪnflu'enʃl/　*adj.* 有很大影响的;有支配力的

interrogator /ɪn'terəgeɪtə/　*n.* 问答机;询问器;质(讯)问者

differentiate /ˌdɪfə'renʃɪeɪt/　*v.* 区分;区别;辨别

rationality /ˌræʃə'næləti/　*n.* 合理性;推理力;理由

biologically /ˌbaɪə'lɒdʒɪkli/　*adv.* 生物学地;从生物学的角度

observable /əb'zɜːvəbl/　*adj.* 能看得到的;能察觉到的

dataset /'deɪtəset/　*n.* 数据集;数据

subfield /'sʌbfiːld/　*n.* 次领域;分场

prediction /prɪ'dɪkʃn/　*n.* 预言;预测;预告

classification /ˌklæsɪfɪ'keɪʃn/　*n.* 分类;分级;类别

descriptor /dɪ'skrɪptə/　*n.* 叙词

consciousness /'kɒnʃəsnəs/　*n.* 意识;观念;看法

superintelligence /ˌsuːpərɪn'telɪdʒəns/　*n.* 超级智能;超级智能体

surpass /sə'pɑːs/　*v.* 超越;超过;优于;胜过

automate /'ɔːtəmeɪt/　*vt.* (使)自动化

scalable /'skeɪləbl/　*adj.* 可攀登的;可称量的

hierarchy /'haɪərɑːki/　*n.* 等级制度(尤指社会或组织);层次体系

supervise /'suːpəvaɪz/　*v.* 监督;管理

ingest /ɪn'dʒest/　*vt.* 摄入;食入;咽下

accessibility /əkˌsesə'bɪləti/　*n.* 可达(及)性;可(易)接近性;可访问性

portfolio /pɔːt'fəʊliəʊ/　*n.* 文件夹;(个人或机构的)投资组合;有价证券组合

milestone /'maɪlˌstəʊn/　*n.* 重要事件;转折点;里程碑

evolution /ˌiːvəˈluːʃn/　　*n.* 进化；演变；发展

coin /kɔɪn/　*vt.* 创造(新词语)；(用金属)铸(币)

perceptron /pəˈseptrɒn/　　*n.* 感知机(模拟大脑识别事物和差异的人工网络)

confine... to...　把……局限于；把……限制于

in conjunction with...　连同……；与……一起

incorporate... into...　将……包括在……内；使……并入……

derive... from...　从……衍生出；起源于……；来自……

trail and error　反复试验(法)(以得出最佳效果)

Exercise

A. Answer the following questions according to the text.

1. According to the text, describe the "Turning Test" briefly.

2. What is the definition of AI offered by John McCarthy in his paper in 2004?

3. How many categories can AI be divided into?

4. What is the relationship among AI, deep learning and machine learning?

5. What are the most common applications of AI system mentioned in the text?

B. Choose the best answer to each of the following questions according to text.

1. _____ is referred to as the "father of computer science".

　A) Stuart Russell　　　　　　　B) Peter Norvig

　C) Alan Turing　　　　　　　　D) John McCarhty

2. In the book, *Artificial Intelligence: A Modern Approach* (1995), four potential goals or definitions of AI were probed into. Which of the following would Alan Turing's definition of AI belong to?

　A) systems that think like humans　　B) systems that act like humans

　C) systems that think rationally　　　D) systems that act rationally

3. Which of the following sentences about AI is NOT true?

　A) It combines computer science and robust datasets.

　B) It includes sub-fields of machine learning and deep learning.

　C) It can help to solve all the difficult problems.

　D) The sub-disciplines of AI consist of AI algorithms.

4. According to the text, all of the following services provided by online virtual agents are true EXCEPT _____.

　A) answering frequently asked questions　B) providing personalized advice

　C) suggesting sizes for users　　　　　　D) optimizing stock portfolios

5. _____ , powered by a deep neural network, beats Lee Sodol, the world champion Go player in a five-game match.

A) Deep Mind's AlphaGo   B) Baidu's Minwa

C) IBM's Deep Blue   D) IBM Waston

 Text B

Facial Recognition

Facial recognition is a way of identifying or **confirming** an individual's identity using their face. Facial recognition systems can be used to identify people in photos, videos, or in **real-time**.

Facial recognition is a category of **biometric** security. Other forms of biometric software include voice recognition, fingerprint recognition, and eye retina or iris recognition.

■ **How does Facial Recognition Work?**

Many people are familiar with face recognition technology through the Face ID used to unlock cellphones (however, this is only one application of face recognition). Typically, facial recognition does not rely on a massive **database** of photos to determine an individual's identity — it simply identifies and recognizes one person as the sole owner of the device, while limiting **access** to others.

Beyond unlocking phones, facial recognition works by matching the faces of people walking past special cameras, to images of people on a watch list. The watch lists can contain pictures of anyone, including people who are not suspected of any wrong doing, and the images can come from anywhere — even from our social media accounts. Facial technology systems can vary, but in general, they tend to operate as follows:

Step 1: Face detection

The camera detects and locates the image of a face, either alone or in a crowd. The image may show the person looking straight ahead or in profile.

Step 2: Face analysis

Next, an image of the face is **captured** and analyzed. Most facial recognition technology relies on 2D rather than 3D images because it can more conveniently match a 2D image with public photos or those in a database. The software reads the **geometry** of your face. Key factors include the distance between your eyes, the depth of your eye **sockets**, the distance from forehead to chin, the shape of your **cheekbones**, and the **contour** of the lips, ears and chin. The aim is to identify the facial **landmarks** that are key to distinguishing your face.

Step 3: Converting the image to data

The face capture process transforms analog information (a face) into a set of digital information (data) based on the person's facial features. Your face's analysis is essentially turned into a **mathematical formula**. The numerical code is called a faceprint. In the same way that **thumbprints** are unique, each person has their own faceprint.

Step 4: Finding a match

Your faceprint is then compared against a database of other known faces. For example, the FBI has access to up to 650 million photos, drawn from various state databases. On Facebook, any photo tagged with a person's name becomes a part of Facebook's database, which may also be used for facial recognition. If your faceprint matches an image in a facial recognition database, then a determination is made.

Of all the biometric measurements, facial recognition is considered the most natural. **Intuitively**, this makes sense, since we typically recognize ourselves and others by looking at faces, rather than thumbprints and irises. It is **estimated** that over half of the world's population is touched by facial recognition technology regularly.

■ How Facial Recognition is Used?

The technology is mostly used for security and law **enforcement**, though there is increasing interest in other areas of use. These include:

Unlocking phones

Various phones, including the most recent iPhones, use face recognition to unlock the device. The technology offers a powerful way to protect personal data and ensures that sensitive data remains **inaccessible** if the phone is stolen. Apple claims that the chance of a **random** face unlocking your phone is about one in 1 million.

Law enforcement

Facial recognition is regularly being used by law enforcement. According to this NBC report, the technology is increasing amongst law enforcement agencies within the US, and the same is true in other countries. Police collects **mugshots** from **arrestees** and compare them against local, state, and federal face recognition databases. Once an arrestee's photo has been taken, their picture will be added to databases to be scanned whenever police carry out another criminal search.

Also, mobile face recognition allows officers to use **smartphones**, tablets, or other **portable** devices to take a photo of a driver or a pedestrian in the field and immediately compare that photo against to one or more face recognition databases to attempt an identification.

Airports and border control

Facial recognition has become a familiar sight at many airports around the world. Increasing numbers of travellers hold biometric passports, which allow them to skip the ordinarily long lines and instead walk through an automated e-Passport control to reach the gate faster. Facial recognition not only reduces waiting times but also allows airports to improve security. The US Department of Homeland Security predicts that facial recognition will be used on 97% of travellers by 2023. As well as at airports and border crossings, the technology is used to enhance security at large-scale events such as the Olympics.

Finding missing persons

Facial recognition can be used to find missing persons and victims of human trafficking. Suppose missing individuals are added to a database. In that case, law enforcement can be alerted as soon as they are recognized by face recognition — whether it is in an airport, retail store, or other public space.

Improving retail experiences

The technology offers the potential to improve retail experiences for customers. For example, **kiosks** in stores could recognize customers, make product suggestions based on their purchase history, and point them in the right direction. "Face pay" technology could allow shoppers to skip long checkout lines with slower payment methods.

Banking

Biometric online banking is another benefit of face recognition. Instead of using one-time passwords, customers can **authorize** transactions by looking at their smartphone or computer. With facial recognition, there are no passwords for hackers to compromise. If hackers steal your photo database, "liveless" detection — a technique used to determine whether the source of a biometric sample is a live human being or a fake representation—should (in theory) prevent them from using it for **impersonation** purposes. Face recognition could make debit cards and signatures a thing of the past.

Healthcare

Hospitals use facial recognition to help with patient care. Healthcare providers are testing the use of facial recognition to access patient records, **streamline** patient **registration**, detect emotion and pain in patients, and even help to identify specific genetic diseases. AiCure has developed an app that uses facial recognition to ensure that people take their medication as **prescribed**. As biometric technology becomes less expensive, adoption within the healthcare sector is expected to increase.

Recognizing drivers

According to this consumer report, car companies are experimenting with facial recognition to replace car keys. The technology would replace the key to access and start the car and remember drivers' preferences for seat and mirror positions and radio station **presets**.

(1132 words)

【Words, Phrases and Expressions】

confirm /kənˈfɜːm/　*v.*（尤指提供证据来）证实；证明

real-time /ˌriːəlˈtaɪm/　*adj.*（计算机处理过程）实时的

biometric /ˌbaɪəʊˈmetrɪk/　*adj.* 生物统计的

database /ˈdeɪtəbeɪs/　*n.*（储存在计算机中的）数据库

access /ˈækses/   *n.* 通道;通路;路径;(使用或见到的)机会,权利

detection /dɪˈtekʃn/   *n.* 侦查;探测;发现;察觉

capture /ˈkæptʃə/   *vt.* 捕获;拍摄;把……输入计算机

geometry /dʒiˈɒmətri/   *n.* 几何学;几何形状;几何结构

socket /ˈsɒkɪt/   *n.* (电源)插座;(电器上的)插孔

cheekbone /ˈtʃiːkbəʊn/   *n.* 颧骨

contour /ˈkɒntʊə/   *n.* 外形;轮廓

landmark /ˈlændmaːk/   *n.* 地标;里程碑

mathematical /ˌmæθəˈmætɪkl/   *adj.* 数学的;(具有)数学(头脑)的;运算能力强的

formula /ˈfɔːmjələ/   *n.* 公式;方程式

intuitively /ɪnˈtjuːɪtɪvli/   *adv.* 直觉地;直观地

thumbprint /ˈθʌmprɪnt/   *n.* 拇指纹印

estimate /ˈestɪmeɪt/   *v.* 估计;估算;估价

enforcement /ɪnˈfɔːsmənt/   *n.* 执行;实施

inaccessible /ˌɪnækˈsesəbl/   *adj.* 无法接近;难以达到的;不可得到的

random /ˈrændəm/   *adj.* 随机的;随意的(非事先决定的或不规则的)   *n.* 随意

mugshot /ˈmʌgʃɒt/   *n.* (警方存档识别罪犯的)面部照片

arrestee /əˈresˈtiː/   *n.* 财产被扣押人

smartphone /ˈsmaːtfəʊn/   *n.* 智能手机

portable /ˈpɔːtəbl/   *adj.* 便携式的;手提的;轻便的   *n.* 手提电脑

kiosk /ˈkiːɒsk/   *n.* (出售报纸、饮料等的)报刊亭;公用电话亭

authorize /ˈɔːθəraɪz/   *vt.* 批准;授权

impersonation /ɪmˌpəːsəˈneɪʃən/   *n.* 模拟;假扮(角色);冒名顶替

streamline /ˈstriːmlaɪn/   *n.* 流线型   *vt.* 使成流线型;使(系统、机构等)效率更高

registration /ˌredʒɪˈstreɪʃn/   *n.* 登记;注册

prescribe /prɪˈskraɪb/   *vt.* 开药方;规定

preset /ˌpriːˈset/   *vt.* 预调;预置;给…预定时间

 Exercise

A. Answer the following questions according to the text.

1. What is "facial recognition" and its main function?

2. What are the categories of biometric security at present?

3. How does facial recognition work?

4. What are the key factors when an image of the face is captured and analyzed?

5. How can hospitals use facial recognition to help with patient care?

B. Read and decide whether they are TRUE(T) or FALSE(F).

(　　)1. Most facial recognition technology relies on 3D images because it can more conveniently match a 3D image with public photos or those in a database.

(　　)2. The face capture process transforms analog information (a face) into a portrait of the person (picture) based on his or her facial features.

(　　)3. The faceprint and thumbprints of each person in the world are special and different from others.

(　　)4. Facial recognition offers the most powerful way to protect personal data and ensures that the phone cannot be unlocked and sensitive data remains inaccessible if it is stolen.

(　　)5. Facial recognition can improve retail experiences for customers and "Face pay" technology could allow them to avoid long checkout lines and pay with faster methods.

Part 3　Extending Skill: Fomat a Formal E-mail

E-mail, an electronic and computer-assisted online communication tool, is often used to transmit virtually every type of correspondence or the daily conduct of work requires, such as simple messages, memos, complex reports, tables of data, graphs and charts, blueprints, pictures. If it can be generated by, scanned into, or downloaded onto a computer, it can be electronically sent through cyberspace to another computer.

In this guide, you will read about writing a formal e-mail and helpful tips on formatting e-mail. Each section provides useful information to assist you in becoming more proficient at using e-mail to communicate in the workplace.

■Format a Formal E-mail

An e-mail functions as both an internal and external method of communication; its three main formatting elements are the heading, the body, and a signature block. The heading of an e-mail usually consists of up to six distinct information fields, including *To, From, Cc, Bcc, Date, Subject, Attached*. Some of these fields are not always visible. The *Bcc* and *Attached* fields, for instance, are visible only when activated by the sender and depending on the e-mail program. The *From* and *Date* fields may not be visible on the sender's template. All the fields are located at the corner of the e-mail template, just below the tool bar. The template itself appears automatically whenever you click on the *New Mail, Reply, Reply to All*, or the *Forward* button found on the tool bar of any e-mail program. Each field of the template is designed to hold specific information as follows.

■Organize the Body of an E-mail

Generally speaking, the body of an e-mail is composed of five parts. E-mail program automatically format the body in single spaced, full block style. It's a design function of the program and meant to ensure that the text of an e-mail appears on the recipient's screen exactly as it does on the sender's.

• Begin with a greeting

The first part of a formal e-mail is what we call the "formal greeting". This is the part of the e-mail where you acknowledge/address the person who will receive the e-mail you are writing. Look at the following examples to start a professional e-mail.

- Dear + Name, ...
- Good morning/afternoon/evening + Name, ...
- Dear Sir/Madam, ...
- To whom it may concern, ...

• Thank the Recipient

The second part of a formal E-mail is where you thank the person you are sending an e-mail to. This part is only included in e-mails that are being sent in response to a previous e-mail that was received. How do we do this? - There are five examples:

- Thank you for contacting us about...(specific)
- Thank you for sending your previous e-mail.
- Thank you for your prompt reply to my previous e-mail.
- I appreciate you getting back to me about...(specific)
- Thanks for your previous e-mail. (more casual)

• State your purpose

The third part of an e-mail is where you state the purpose or reason for sending the e-mail. This part is the core of the e-mail, but it should also be clear and easy to understand. There are three different ways to state your purpose here.

①5W's Method: To give the background details and description of the purpose.

- Who is involved?
- What is being done?
- When is it being done?
- Where is it being done?
- Why is it being done?

Example

My team and I have been working hard to fix your computer issues. We initially anticipated that it would only take us a few days to rectify the issues. However, after doing some more research, we realized that it will take us another week to fully repair the broken parts. All of the repairs are being taken care of in-house and I am overseeing each process. I know that you need your computer quickly, so we are working hard to get everything right. We truly appreciate your patience.

②**3 Points Method:** To give the main ideas, details, or parts related to your purpose. Depending on the e-mail, you may need more than three points.

- first important point ＋ detail/reason

- second important point ＋ detail/reason

- third important point ＋ detail/reason

- ...

Example

Assessment &. Work needed

We have conducted an overall assessment of your computer and determined that you need a new motherboard.

Timeline

The estimated time to complete this project is 3 – 5 days.

Cost

The parts will be a total of ＄500 and the labor will be ＄150.

So, in total, the cost will be ＄650.

③**List Method:** To visually show clearly the purpose of your E-mail.

- Introduce/Explain the list

- List out each step/part of the process

- Sum up the list

Example

Below you will find a list of all of the things my team and I found wrong with your computer:

a. Damaged hard drive (＊repairable)

b. Outdated video card (＊needs to be replaced)

c. Corrupted motherboard (＊needs to be replaced)

In our opinion, these are the three main components of your computer that need immediate attention.

- **Add closing remarks**

The fourth part of a formal e-mail is where you bring your e-mail to a close. This part is basically the wrap-up and last comment. It can also help the recipient know what you would like to happen next. There are five ways to do this.

- Thank you again for your time and cooperation.

- Thank you for your work thus far on this project.

- Thanks again and if you have any questions, please feel free to contact me anytime.

- If you have any questions or concerns, don't hesitate to let me know.

- I look forward to hearing from you soon.

- **End with a closing**

The fifth part of a formal e-mail is the final "goodbye" of your e-mail. This is the part where you include an appropriate closing word or phrase and then add your name. Here we have five simple ways to end.

- Sincerely,

  (name)

- Regards,

  (name)

- Best regards,

  (name)

- Have a great day,

  (name)

- Respectfully,

  (name)

■ **Fill in the Signature Block**

The signature block in an e-mail supplies the contact information belonging to the sender. This is the last item in an e-mail. It is always located on the left hand margin below the signature line. A signature block should contain all the contact information a recipient might require in order to respond to an e-mail. It should begin with the sender's name, title and business organization. A physical location, phone numbers, e-mail address, and Web site should follow.

【Exercise】

Please write an e-mail to your supervisor about the work progress of your graduation thesis or project in the past month.

Part 4　Optional Exercises

A. Translate the following paragraph into English.

　　人工智能是计算机科学的一个分支,二十世纪七十年代以来被称为世界三大尖端技术之一(空间技术、能源技术、人工智能)。近些年来,它获得了迅速的发展,在很多学科领域都获得了广泛的应用,并取得了丰硕的成果。人工智能已逐步成为一个独立的分支,无论在理论和实践上都已自成一个系统。它的研究成果正在逐渐融入人们的生活中,为人类创造更多的幸福。

B. Translate the following paragraph into Chinese.

Many people are familiar with face recognition technology through the Face ID used to unlock cellphones (however, this is only one application of face recognition). Typically, facial recognition does not rely on a massive database of photos to determine an individual's identity — it simply identifies and recognizes one person as the sole owner of the device, while limiting access to others. Beyond unlocking phones, facial recognition works by matching the faces of people walking past special cameras, to images of people on a watch list.

C. Critical thinking. Sit in groups of three or four and discuss the following topic.

The term "artificial intelligence" was put forward by John McCarthy at the first-ever AI conference at Dartmouth College in 1956. Since then, it has not only had a profound impact on Natural Science but also brought infinite convenience to our life. Therefore, some people hope to develop AI more vigorously and make our life fully intelligent in the near future. Others, by contrast, are worried that we are not prepared for a world in which computers are more intelligent than humans and that we will be surpassed even replaced by AI. Discuss both of these views and give your own opinion.

Unit 10

Virtual Reality

Part 1   Vocabulary

A. Choose the explanations in Column B that best match the terms in Column A.

Column A	Column B
1. Virtual Reality	a. an environment created by computer
2. three-dimensional	b. having, or appearing to have, length, width and depth
3. computer-generated environment	c. a hardware that is held on the head by one or more bands or straps
4. omnidirectional treadmills	d. a device that detects and responds to a signal or stimulus of movements
5. Augmented Reality	e. a device that controls or has power or authority to control by radio waves rather than over wires
6. motion sensors	f. the real-time use of information in the form of text, graphics, audio and other virtual enhancements integrated with real-world objects
7. trackers	g. images created by a computer that appear to surround the person looking at them and seem almost real
8. Goggles	h. a device that finds other people or animals by following the marks left by their feet and other signs that show where they have been.
9. wireless controllers	i. an electronic apparatus that covers the eyes and is used to enhance vision (as at night) or to produce images (as of a virtual reality)
10. headset	j. a piece of equipment consisting of a wheel with steps around its edge or a continuous moving belt. The weight of a person or animal walking on it causes the wheel or belt to turn in all directions.

B. Use the words from Exercise A to label the following pictures.

1. _____

2. _____

3. _____

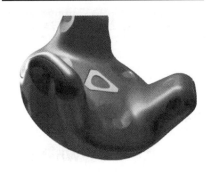

4. _____

5. _____

6. _____

C. Complete the following sentences with the words given below. Change the form if necessary.

technical	immerse	sensory	peripheral	mediate
entail	emulation	illusion	synchronize	interface
virtual	disrupt	processing	mechanism	substantively

1. New technology has enabled development of an online "_____ library".

2. Then the _____ experiments show that this method is effective and practicable.

3. Conscious mind is our awake mind, which is directly linked with our all _____ organs.

4. The Data _____ analysis is likely to prevail in the long run.

5. Further development in techniques and technologies may _____ the need for future modification of this standard.

6. Children _____ themselves in their own artificial worlds of miniatures.

7. The idea of absolute personal freedom is an _____.

8. The sound track did not _____ with the action.

9. The new system _____ with existing telephone equipment.

10. Workstation and _____ devices are directly connected to central computer.

Part 2　Reading

 Text A

What is Virtual Reality?

The definition of "**virtual** reality" comes, naturally, from the definitions for both "virtual" and "reality". The definition of "virtual" is "near" and "reality" is "what we experience as human beings". So the term "virtual reality" basically means "near-reality". This could, of course, mean anything but it usually refers to a specific type of reality **emulation**.

We know the world through our senses and perception systems. In school we all learned that we have five senses: taste, touch, smell, sight and hearing. These are however only our most obvious sense organs. The truth is that humans have many more senses than this, such as a sense of balance for example. These other **sensory** inputs, plus some special **processing** of sensory information by our brains ensures that we have a rich flow of information from the environment to our minds.

Everything that we know about our reality comes by way of our senses. In other words, our entire experience of reality is simply a combination of sensory information and our brains sense-making **mechanisms** for that information. It stands to reason then, that if you can present your senses with made-up information, your perception of reality would also change in response to it. You would be presented with a version of reality that isn't really there, but from your perspective it would be perceived as real. Something we would refer to as a virtual reality.

So, in summary, virtual reality **entails** presenting our senses with a computer-generated virtual environment that we can explore in some **fashion**.

Answering "what is virtual reality" in technical terms is straight-forward. Virtual reality is the term used to describe a three-dimensional, computer-generated environment which can be explored and interacted with by a person. That person becomes part of this virtual world or is **immersed** within this environment and whilst there, is able to manipulate objects or perform a series of actions.

Although we talk about a few historical early forms of virtual reality elsewhere on the site, today virtual reality is usually implemented using computer technology. There are a range of systems that are used for this purpose, such as headsets, omni-directional treadmills and special gloves. These are used to actually stimulate our senses together in order to create the **illusion** of reality.

This is more difficult than it sounds, since our senses and brains are evolved to provide us with a finely **synchronized** and **mediated** experience. If anything is even a little off, we can usually tell. This is where you'll hear terms such as "immersiveness" and "realism" enter the conversation. These issues that divide **convincing** or enjoyable virtual reality experiences from **jarring** or unpleasant ones are partly **technical** and partly **conceptual**. Virtual reality technology needs to **take** our **physiology into account**. For example, the human visual field does not look like a video **frame**. We have (more or less) 180 degrees of vision and although you are not always consciously aware of your **peripheral** vision, if it were gone, you'd notice. Similarly, when what your eyes and the **vestibular** system in your ears tell you are in conflict it can cause motion sickness, which is what happens to some people on boats or when they read while in a car.

If an implementation of virtual reality manages to get the combination of hardware, software and sensory synchronicity just right it achieves something known as a sense of presence. Where the subject really feels like they are present in that environment.

This may seem like a lot of effort, and it is! What makes the development of virtual reality worthwhile? The potential entertainment value is clear. Immersive films and video games are good examples. The entertainment industry is after all a multi-billion dollar one and consumers are always keen on **novelty**. Virtual reality has many other, more serious, applications as well.

There are a wide variety of applications for virtual reality which include architecture, sport, medicine, the arts and entertainment. Virtual reality can lead to new and exciting discoveries in these areas which impact upon our day-to-day lives.

Wherever it is too dangerous, expensive or impractical to do something in reality, virtual reality is the answer. From trainee fighter pilots to medical application's trainee surgeons, virtual reality allows us to take virtual risks in order to gain real world experience. As the cost of virtual reality goes down and it becomes more mainstream you can expect more serious uses, such as education or productivity applications, to **come to the fore**. Virtual reality and its cousin augmented reality could **substantively** change the way we **interface** with our digital technologies, continuing the trend of **humanising** our technology.

There are many different types of virtual reality systems but they all share the same characteristics such as the ability to allow the person to view three-dimensional images. These images appear **life-sized** to the person.

Plus, they change as the person moves around their environment which **corresponds with** the change in their field of vision. The aim is for a **seamless** join between the person's head and eye movements and the appropriate response, e. g. change in perception. This ensures that the virtual environment is both realistic and enjoyable.

A virtual environment should provide the appropriate responses—in real time—as the person explores their surroundings. The problems arise when there is a delay between the person's actions and system response or **latency** which then **disrupts** their experience. The person becomes aware that they are in an artificial environment and adjusts their behavior accordingly which results in a **stilted**, mechanical form of interaction. The aim is for a natural, free-flowing form of interaction which will result in a memorable experience.

Virtual reality is the creation of a virtual environment presented to our senses in such a way that we experience it as if we were really there. It uses a host of technologies to achieve this goal and is a technically complex **feat** that has to **account for** our perception and cognition. It has both entertainment and serious uses. The technology is becoming cheaper and more widespread. We can expect to see many more innovative uses for the technology in the future and perhaps a fundamental way in which we communicate and work thanks to the possibilities of virtual reality.

(1,044 words)

【Words, Phrases and Expressions】

virtual /ˈvəːtʃuəl/　*adj.* 实质上的,事实上的;(计算机)虚拟的

emulation /ˌemjʊˈleɪʃn/　*n.* 仿效,仿真;竞争

sensory /ˈsensəri/　*adj.* 感觉的,感受的,感官的;传递感觉的

processing /ˈprəʊsesɪŋ/　*n.* (数据)处理;整理;配置;工艺(生产方法)设计

mechanism /ˈmekənɪzəm/　*n.* [生] 机制,机能;[乐] 机理;(机械)结构;机械装置,(故事的)结构;[艺] 手法;技巧;途径

entail /ɪnˈteɪl/　*v.* 牵涉;使必要;势必造成

fashion /ˈfæʃn/　*n.* 时尚(界);时装;以……方式　*v.* 制作,塑造

immerse /ɪˈməːs/　*vt.* 浸没;施浸礼;沉迷……中,陷入

illusion /ɪˈluːʒn/　*n.* 错觉;幻想;错误观念;假象

synchronize /ˈsɪŋkrənaɪz/　*vt.* 使同步;使同时　*vi.* 同时发生;共同行动

mediate /ˈmiːdɪeɪt/　*vt.* 经调解解决;斡旋促成　*vi.* 调停,调解,斡旋

convincing /kənˈvɪnsɪŋ/　*adj.* 人信服的,有说服力的

jarring /ˈdʒɑːrɪŋ/　*adj.* 刺耳的,不和谐的　*n.* 震动,冲突

technical /ˈteknɪkl/　*adj.* 技术的,技能的;技艺的,技巧的;专业的

conceptual /kənˈseptʃuəl/   *adj.* 观念的,概念的

physiology /ˌfɪzɪˈɒlədʒɪ/   *n.* 生理学;生理机能

frame /freɪm/   *n.* 框架,构架;体系;眼镜框   *v.* 给……装/做框;表达;制定;诬陷

peripheral /pəˈrɪfərl/   *adj.* 外围的;次要的;(神经)末梢区域的 n. 外部设备

vestibular /vesˈtɪbjʊlə/   *adj.* 门厅的,门口走廊的,前庭的

novelty /ˈnɒvltɪ/   *n.* 新奇;新奇的事物;新颖小巧而价廉的物品
                   adj. 新奇的;风格独特的

substantively /ˈsʌbstəntɪvlɪ/   *adv.* 真实地;实质上

interface /ˈɪntəfeɪs/   *n.* 界面;[计]接口;交界面
                   v. (使通过界面或接口)接合,连接;[计算机]使联系
                   vi. 相互作用(或影响);交流,交谈

humanise /ˈhjuːmənaɪz/   *v.* 赋予人性;使通人情;变为有人性;变为有情

life-sized/ˌlaɪfˈsaɪzd/   *adj.* (艺术作品)与真人(实物)一样大的

seamless /ˈsiːmləs/   *adj.* 无缝的;无漏洞的

latency /ˈleɪtənsɪ/   *n.* 潜伏;潜在因素

disrupt /dɪsˈrʌpt/   *v.* 扰乱,使中断

stilted /ˈstɪltɪd/   *adj.* (动作或言语)生硬的,不自然的

feat /fiːt/   *n.* 功绩,伟业;技艺

take...into account   考虑到……

come to the fore   走到前面来;发作;涌现出来;惹人注意

correspond with   与……通信,与……相一致;切合

account for   说明(原因、理由等);导致,引起;(在数量、比例上)占;对……负责

 Exercise

A. Answer the following questions according to the text.

1. What does the term "virtual reality" basically mean?

2. What is virtual reality in technical terms?

3. When will the motion sickness be caused?

4. What areas can virtual reality be applied to?

5. What kind of problems may arise as the person explores in the virtual environment?

B. Choose the best answer to each of the following questions according to the text.

1. We get to know about our reality through _____.

   A) reading   B) senses   C) imagination   D) thinking

2. Which of the following does not belong to our most obvious sense organs?

A) taste B) touch C) sight D) sense of balance

3. Which of the following are used to create virtual reality today?

   A) headsets B) omni-directional treadmills

   C) special gloves D) All of the above.

4. What is the characteristic do many different types of virtual reality systems share?

   A) The ability to allow the person to explore the surroundings.

   B) The ability to allow the person to be immersed in the environment.

   C) The ability to allow the person to view three-dimensional images.

   D) The ability to allow the person to perform a series of actions.

5. Which of the following statements is not true?

   A) Although we are not constantly conscious of our peripheral vision, we would notice if it were gone.

   B) We cannot expect to see many more innovative uses for technology in the future because it is becoming more and more expensive.

   C) Virtual reality can be a solution when you do something dangerous or impractical.

   D) A sense of presence can be achieved if the combination of hardware, software and sensory synchronicity gets right in the implementation of virtual reality.

 Text B

Virtual Reality in Education

Would you rather read about the moon landing or see for yourself what it was like to walk on the moon with Neil Armstrong and Buzz Aldrin? Believe it or not, experiencing the latter is just as possible as the former, thanks to the rise in virtual reality (VR).

The days of learning being **restricted solely** to reading textbooks and listening to boring lectures are numbered, and when they're gone, students won't miss them. Research shows that textbooks don't generally improve student achievement and traditional stand-and-deliver lectures in universities lead to higher student failure rates than active learning methods.

And while there are plenty of active learning **techniques** to choose from, including simply asking students questions or arranging students for group work, more and more educators are seeing VR's true potential. According to a recent survey of teachers and students, 90% of educators believe VR may help increase student learning. Perhaps more importantly, the survey also found that 97% of students would attend a class or course with VR, which could significantly decrease **dropout** rates.

It comes as no surprise, then, that education is one of the largest **sectors** for VR investment. Indeed, industry forecasts predict that VR in education will be a $700 million industry by 2025. But is VR in education all that it's **cracked up to be**? Let's find out.

Virtual reality is a computer-generated environment that creates the immersive illusion that the user is somewhere else.

Instead of looking at a screen in front of them, VR allows people to interact with an **artificial** three-dimensional environment through electronic devices that send and receive information like motion sensors and movement trackers.

The most essential VR device is the headset, which generally looks like a pair of thick goggles. **Fitted out with** a unique screen and motion sensors, a VR headset tracks the user's movement and changes the angle of the screen accordingly. Optional **accessories** can enhance user experience and include things like:

Hand gloves. Wireless controllers that capture full hand and finger action in virtual reality and provide the user the **sensation** of touch.

Treadmills. A mechanical device that looks nothing like the gym equipment you're used to, a VR treadmill translates your real-life body movements into virtual motion.

Vive Trackers. Small hockey **puck-esque** devices that bring physical objects you own into the virtual world.

Below are just a few examples of how students and educators use VR at all education levels, including K12 education, special education, higher education, and vocational training.

K12 education

At the K12 level (kindergarten to 12th grade in the US), virtual field trips are among the most common ways educators use VR. For example, in 2019, the Schaumburg School District 54 in Illinois **utilized** virtual reality **kits** in each of its 28 schools to bring students on virtual field trips to the moon, World War I battlefields, and the Great Hall at Ellis Island.

The enthusiasm from kids has been overwhelming, said Associate Superintendent Nick Myers in an interview with *EdTech* magazine. "We've seen truly emotional reactions to it because the students can see it, they can navigate through and be part of the experience they're learning about."

VR field trips are becoming so popular in education because, in addition to providing immersive and engaging experiences, they're also accessible. Not every student may be able to join their classmates for a real-world trip to a museum or another country, whether because of a disability or expense. With VR, every student can go on the same trip at no cost. Because they don't require expensive transport and **logistics**, virtual field trips are more cost-effective for schools.

Other uses of VR in K12 education include language immersion and virtual lab **simulation**. Language immersion allows students to connect with people all over the world. On the other hand, virtual lab simulation gives STEM students the option of experimenting in million-dollar labs or mixing different chemicals in a virtual chemistry class without fear of **blowing** anything **up** in real life.

Special education

For students with special needs, VR creates new opportunities to safely explore the world and practice real-world skills, like obeying traffic signals or interacting with police officers, in a no-risk environment.

For example, Danvers Public Schools district in Massachusetts used VR to introduce new students to the district's middle school building in advance, something that was particularly helpful for students with disabilities.

Higher education

Choosing the right university can be a **daunting** and exhausting experience. With VR, applicants can go on virtual reality campus tours to see what it would be like to attend a college or university in another city or even another country.

For example, the University of Michigan athletic department uses VR technology to give potential **recruits** the chance to see and feel the campus and the athletic facilities from wherever in the world they may be.

But with VR, you may not even have to attend a physical university. During the past several years, Steven Hill, professor at the University of North Carolina at Chapel Hill, **ditched** Zoom lectures for a virtual 3D version of his classroom. Students can walk around the classroom, talk to each other at different gathering spaces, and even break into groups.

Of course, VR is useful for learners who attend physical institutions, as well. At the Beijing University of Chinese Medicine, students use VR to learn **acupuncture**. In the UK, the University of Westminster has implemented a virtual training center that allows criminal law students to investigate potential murder scenes.

Vocational training

Unfortunately, vocational training is often seen as a second choice — something that students do when they can't get into a university. Some trade schools are trying to change this by using VR technology to give **prospective** students a glimpse into a vocational graduate's daily life.

In addition to attracting new students to trade schools, VR can also give trainees more opportunities to practice essential skills in a safe environment. For example, electricians can **rewire** a house with fewer safety hazards. Moreover, because trainees work with virtual materials, trade schools can save **tons of** money on physical materials.

According to one study that looked at 1,000 students in three universities, the implementation of VR in classrooms led to students improving **by a full letter grade**. One of the main advantages of using VR in education is that it raises students' grades.

In another instance, a hospital found that using VR to train medical students increased their **retention** rate by 80% a year after the lecture compared to 20% a week after when they didn't use VR. This boost in retention isn't so surprising when you consider that VR promotes student curiosity and keeps them **engaged** even when learning challenging topics.

　　Other benefits of virtual reality include increased collaboration, cultural competence, and fewer distractions. VR can also help students build better habits. Indeed, according to recent research, after using VR, people have been found to exercise more as well as show more **empathy**, among other things.　　(1,164 words)

【Words, Phrases and Expressions】

restrict /rɪˈstrɪkt/　*vt.*（以法规）限制；限定（数量、范围等）；束缚；妨碍

solely /ˈsəʊli/　*adv.* 唯一地；仅；只；唯；单独地

technique /tekˈniːk/　*n.* 技巧；技艺；工艺；技术；技能

dropout /ˈdrɒpaʊt/　*n.* 辍学者；退学者；拒绝传统社会的人

sector /ˈsektə/　*n.* 部门；领域；行业；区域；地带

artificial /ˌɑːtɪˈfɪʃl/　*adj.* 人造的；人工的；假的；非自然的；虚假的

accessory /əkˈsesəri/　*n.* 附件；配件；附属物；（衣服的）配饰

sensation /senˈseɪʃn/　*n.* 感觉；知觉；感觉能力；知觉能力；直觉

puck-esque /pʌk-esk/　*adj.* 冰球式的

utilize /ˈjuːtəlaɪz/　*vt.* 使用；利用；运用；应用

kit /kɪt/　*n.* 配套元件；成套工具；成套设备；全套衣服及装备

logistics /ləˈdʒɪstɪks/　*n.* 后勤；物流；组织工作

simulation /ˌsɪmjuˈleɪʃn/　*n.* 模拟；仿真；假装；冒充

daunting /ˈdɔːntɪŋ/　*adj.* 令人生畏的；使人畏惧的；让人气馁的

recruit /rɪˈkruːt/　*n.* 新兵；新警员；新成员；新生

ditch /dɪtʃ/　*v.* 摆脱；抛弃；丢弃

acupuncture /ˈækjupʌŋktʃə/　*n.* 针灸；针刺疗法

prospective /prəˈspektɪv/　*adj.* 预期的；潜在的；有望的；可能的；即将发生的

rewire /ˌriːˈwaɪə(r)/　*vt.* 给（建筑物或设备）换新电线

retention /rɪˈtenʃn/　*n.* 保持；维持；保留

engage /ɪnˈgeɪdʒ/　*v.* 从事；聘用；吸引住（注意力、兴趣）

empathy /ˈempəθi/　*n.* 移情；同情；共鸣；同感

cracked up to be　被吹捧

fit out with　装备；供给……必要之物

blow... up　炸毁

tons of　大量的，很多的

by a full letter grade　整整一个档次

Exercise

A. Answer the following questions according to the text.

1. What does the research show about the textbooks reading and traditional stand-and-deliver lectures?

2. What was discovered in terms of educators and students according to the survey of teachers and students?

3. What are the optional VR accessories that can enhance user experience?

4. From which education levels does the author give examples to illustrate the application of VR in education?

5. What advantages of using VR in education are mentioned in the text?

B. Read the following statements and decide whether they are TRUE(T) or FALSE(F).

(   )1. It is not surprising that education is one of the largest fields for VR investment.

(   )2. VR allows people to interact with the screen in front of them through electronic devices that send and receive information.

(   )3. The headset, which looks like a pair of goggles in general, is the most essential VR device.

(   )4. Virtual lab simulation enables STEM students to conduct experiments in million-dollar labs or mix different chemicals in a virtual chemistry class without fear of ruining anything in real life.

(   )5. During the past several years, a professor ditched a virtual 3D version of his classroom in which students can walk around the classroom, talk to each other at different gathering spaces.

Part 3   Extending Skill: Writing a Resume

A resume is the key job search tool and should be limited to one page or two pages if you have substantial experience, make sure that it is well organized, well designed, easy to read, and free of errors. Before writing, it is necessary to analyze your background, deciding what information would be most important and related to that kind of job. List the following:

• Schools attended, degrees, major field of study, academic honors, grade point average (if recently graduated), particular and relevant academic projects

• Jobs held, primary and secondary duties in each of them when and how long you held each job, promotions, skills you developed in your jobs

• Other experiences and skills you have developed that would be of value; extracurricular activities that have contributed to your learning experience; leadership. interpersonal, and communication skills you have developed; any collaborative work you have performed; computer skills you have acquired.

Most resumes have information arranged chronologically in the following categories:

- Heading (name and contact information)
- Job Objective (optional)
- Education
- Employment Experience.
- Computer Skills (optional)
- Honors and Activities
- References

If you are recent graduate without much work experience, list education first. If you have many years of job experience, list employment experience first. In your education and employment sections, use a chronological sequence with most recent experience first.

• **Heading**

At the top of your resume, include all of your contact information including: name, address, telephone, fax, and e-mail address.

• **Objective**

A job objective introduces the material in a resume and helps the reader quickly understand your goal. Write your objective in three lines or less. For example: a full-time software programme analyst aimed at solving engineering problems.

• **Education**

List the school(s) you have attended, the degrees you received and the dates you received then your major field(s) of study, and any academic honor you have earned. Include your grade point average if it is high. List courses if they are unusually impressive or if your resume is otherwise sparse.

• **Employment experience**

You can organize your employment experience chronologically, starting with your most recent job,

- Include jobs or internships when they relate directly to the position, provide a concise description of your primary and secondary responsibilities.

- Include extracurricular experiences, especially those involving a leadership position or community-service.

- Use action verbs with precision and conciseness. Do not use "I".

• **Computer skills**

If you are applying for a job that requires computer knowledge, include specific languages, software, and hardware you are familiar with.

• **Honors and activities**

If there is still room on the resume (less than two pages), select items such as fluency in foreign languages, specialized technical knowledge, student or community activities, professional or club memberships, and published works.

【Exercise】

Write a resume based on the following information.

You are Li Hua, an undergraduate in grade four, you want to apply for a software engineer position in the R&D center, a financial company.

Part 4　Optional Exercises

A. Translate the following paragraph into Chinese.

VR field trips are becoming so popular in education because, in addition to providing immersive and engaging experiences, they're also accessible. Not every student may be able to join their classmates for a real-world trip to a museum or another country, whether because of a disability or expense. With VR, every student can go on the same trip at no cost. Because they don't require expensive transport and logistics, virtual field trips are more cost-effective for schools.

B. Translate the following paragraph into English.

虚拟现实是一种完全数字化的体验,既可以模拟现实世界,也可以建立完全不同于现实的世界。虚拟现实一词是指计算机生成的三维环境。为了体验虚拟现实并与之互动,您需要合适的设备,例如一副 VR 眼镜或耳机。虚拟现实技术用于创造身临其境的体验,除了流行的游戏用例之外,虚拟现实还应用于各种行业,例如医学、建筑、军事等。

C. Write an essay on the topic "The Advantages and Disadvantages of Virtual Reality". Your essay may cover the points given below.

a. The advantages of virtual reality.

b. The disadvantages of virtual reality.

c. Your views.